Monitoring of Soil-Structure Interaction

Join Us on the Internet

WWW: http://www.thomson.com
EMAIL: findit@kiosk.thomson.com

thomson.com is the on-line portal for the products, services and resources available from International Thomson Publishing (ITP).

This Internet kiosk gives users immediate access to more than 34 ITP publishers and over 20,000 products. Through *thomson.com* Internet users can search catalogs, examine subject-specific resource centers and subscribe to electronic discussion lists. You can purchase ITP products from your local bookseller, or directly through *thomson.com.*

Visit Chapman & Hall's Internet Resource Center for information on our new publications, links to useful sites on the World Wide Web and an opportunity to join our e-mail mailing list. Point your browser to: **http://www.chaphall.com** or **http://www.thomson.com/chaphall/civeng.html** for Civil Engineering

A service of ITP®

Monitoring of Soil-Structure Interaction

Instruments for Measuring Soil Pressures

George E. Lazebnik
Section Head
Energoproect Institute
Kiev, Ukraine

Translated and Edited by
Gregory P. Tsinker, Ph.D., P.E.
President
Tsinker & Associates, Inc.
Ontario, Canada

SPRINGER-SCIENCE+BUSINESS MEDIA, B.V.

Cover design: Curtis Tow Graphics

Originally published by Chapman & Hall in 1998
Softcover reprint of the hardcover 1st edition 1998

1 2 3 4 5 6 7 8 9 10 XXX 01 00 99 98

Library of Congress Cataloging-in-Publication Data

Lazebnik, George E.
Monitoring of soil-structure interaction : instruments for measuring soil pressures / George E. Lazebnik ; translated and edited by Gregory P. Tsinker.
p. cm.
Includes bibliographical references (p.) and index.
ISBN 978-1-4613-7740-5 ISBN 978-1-4615-5979-5 (eBook)
DOI 10.1007/978-1-4615-5979-5
1. Soil-structure interaction. 2. Earth pressure--Measurement. 3. Foundations. I. Tsinker, Gregory P. II. Title.
TA711.5.L39 1997
624.1'51--DC21 97-8343
CIP

British Library Cataloguing in Publication Data available

"Monitoring of Soil-Structure Interaction" is intended to present technically accurate and authoritative information from highly regarded sources. The publisher, editors, authors, and contributors have made every reasonable effort to ensure the accuracy of the information, but cannot assume responsibility for the accuracy of all information, or for the consequences of its use.

To the memory of my mother, Sophy V. Lazebnik, whose humanism, inexhaustible energy, selflessness, and rare gentleness have always served me as a source of strength and inspiration.

Table of Contents

Preface **xi**

Introduction **xiii**

1 Soil Pressure Measurement Methods: Measurement Specifics and Instruments **1**

1.1 Direct and Indirect Methods 1
1.2 Some Specifics of Soil Pressure Measurement 2
1.3 Hydraulic Multiplicator 6
1.4 Orientation Effect on Cell Performance and Design 9
1.5 Cell for Measuring Both Static and Dynamic Soil Pressures 19

2 Contact Soil Pressure Measurement Techniques **23**

2.1 Conventional Approach and Experimental Verification 23
2.2 Stress Deviator and Its Effects on Variation in Soil Pressure Cell Readings 30

3 Calibration Apparatuses and Calibration Methods of Soil Pressure Cells **38**

3.1 State-of-the-Art Review 38
3.2 Improved Calibration Devices and Investigation of Their Performance 45

4 Stiff Contact Pressure Cells **57**

4.1 Stress-Strain Measuring Systems: General 57
4.2 Developing of Contact Pressure Cells 59
4.3 Contact Pressure Cell With Vibrating Wire Type Transducer for Mass Production 70
4.4 Effects of Cell Size on Scatter in Results of Soil Pressure Measurement 73

5 Errors in Measuring Soil Contact Pressures by Means of Nonstiff Pressure Cells **79**

5.1 Introduction 79
5.2 Instruments and Devices Used in Experiments for Determining the Errors in Soil Pressure Measurements by Nonstiff Cells 80

5.3 Changes in Properties of Cohesionless Soils That Occur Due to Effects of Compression 83
5.4 Effect of Distance Between Adjacent Soil Pressure Cells on Instrument Performance 88
5.5 Effect of Depth of Soil Above Instrument's Pressure-Bearing Membrane on Instrument Performance 89
5.6 Errors in Soil Pressure Measurements by Nonstiff Soil Pressure Cells 92
5.7 Soil Stress State Investigation at Soil Pressure Cell Contact Surface 99
5.8 Results of Investigation of Systematic Relative Error in Stress Measurements in Soil Performed with Nonstiff Instruments 106

6 Full-Scale and Field Experiments: Monitoring of In Situ Soil/Structure Interaction 112
6.1 Arrangement (Layout) of Pressure Cells and Installation Procedures 112
6.2 Effects of Misalignment of Cell Contact Surface with Surface of the Structure 118
6.3 Influence of Size of Soil Grains on Cell Performance 121
6.4 Interaction Between Adjacent Pressure Cells in a Mass of Soil 123

7 Contact Stresses at Interface Between a Stiff Foundation Block and Foundation Soil: Large-Scale Model Tests 126
7.1 Round Stiff Concrete Block 1.1 m in Diameter 127
7.2 Round Stiff Concrete Block 2.2 m in Diameter 128
7.3 Stiff Rectangular Concrete Block 3 × 1 × 0.6 m 128

8 Field Investigation. Case Histories 135
8.1 Round Concrete Foundation for Cylindrical Concrete Shaft to Support a Cable Roof System 135
8.2 Cellular Square Footing Constructed on Granular Foundation Soils 138
8.3 Cellular Square Footing Constructed on Hard Foundation Soil 141
8.4 Foundation of a Tall Building 144
8.5 Foundation Slab to Support Loads from Turbine and 300-MW Generator 148

9 Stiff Foundations on Cohesive and Nonhomogeneous Soils 153
9.1 Perforated Stiff Foundation System on Cohesive Soils 153
9.2 Solid Concrete Slab Constructed Upon a Nonhomogeneous Foundation Soil 154
9.3 Soil Pressure from a Water Reservoir 158
9.4 Influence of Nonuniform Densification on Shape of Soil Reaction Diagram 160
9.4.1 Stiff Circular Concrete Block 2.2 m in Diameter 161
9.4.2 Stiff Rectangular Concrete Block 3 × 1 × 0.6 m 162

10 Lateral Earth Pressure "At Rest" 165
10.1 Introduction 165
10.2 Experimental Determination of K_0 in a Granular Soil Mass 166
10.3 Coefficient of Lateral Soil Pressure, K, at Interfaces with Stiff Walls 170
10.4 Comparison of K_0 with K and Proposed Formula for Determining K_0 183

11 Soil Pressure on Retaining Walls **184**
11.1 Introduction 184
11.2 Large-Scale Investigation of a Single-Anchored Sheetpile Wall 185
11.3 Soil Pressure Against Nonyielding Retaining Wall 198

12 Vertical Distribution of Stresses Within Soil Mass at the Central Part of a Stiff Slab **204**
12.1 Stiff Rectangular Concrete Block 1.8 × 0.6 × 0.7 m 204
12.2 Stiff Rectangular Concrete Block 2.85 × 0.7 × 0.7 m 206
12.3 Stiff Round Concrete Block 1.1 m in Diameter 206
12.4 Conclusions 207

References **209**

Index **219**

Preface

As a practicing engineer and researcher I have dedicated a great deal of my longer than 40-year professional career to the development, manufacture, and practical implementation of instruments for taking measurements on soil stresses at soil/structure interfaces, and within a soil mass at some distance from a foundation's underside. My experience with instruments has been obtained in both large-scale laboratory tests and in field investigations. The latter are usually associated with a number of subtle and severe test conditions, and with a stringent demand for accuracy and stability in the performance of stress measuring sensors and stress/strain measuring instrumentation, particularly where measurements are to be taken over long periods of time to monitor the performance of the important structures.

This book, however, is not a summary of the state of the art of this subject matter. This has been done earlier in a comprehensive work by Dunnicliff (1988) and has more recently been discussed in Bhatia and Blaney (1991). In this work I have attempted to describe in a concise form the equipment, methods, and results of some important experiments I have been involved with for more than 30 years while serving as a senior scientist at the Ukrainian Branch of the Research Institute for Investigation of Structures (Russian abbreviation: NIISK) and as head of the specialized group responsible for field monitoring of the soil/structure interaction of miscellaneous important structures (e.g., foundations of nuclear power stations, large buildings, etc.) at the State Institute "Energoproect" (formerly "Atomenergoproect") all in Kiev, Ukraine.

I have also attempted to briefly describe results of some important investigations on soil/structure interaction that have been carried out in the past 30 to 40 years in the republics of the former USSR. The results of these investigations are published in a great number of local papers and books. However, despite the significant achievements of my countrymen in this field of science, the results of their work are largely unknown beyond the borders of the former USSR.

Certainly this has occurred because of decades of my country's isolation from the West, but it is also due to language barriers. References to these works are made elsewhere in this book.

This work is intended to be a useful source of information for practitioners involved with instrument manufacturing, contractors specializing in implementation of geotechnical instrumentation programs for investigation and monitoring of the phenomenon of soil/structure interaction, and individuals involved with laboratory investigations on soil/structure interaction. This book may also be of special value to those involved with mathematical modeling of soil/structure interaction phenomenon; the results of the large-scale model tests as well as the full-scale investigations presented in this book can be used for the calibration of mathematical models. I believe this book will also help faculty members and postgraduate students better understand the different aspects of instruments used for measurements of soil pressures at soil/structure interfaces and within a soil mass.

This book has been several years in preparation and was written with the help of many people. First of all, I would like to thank all my colleagues at NIISK, "Energoproect," and associated organizations with whom I have worked for many years on many important projects, some of which are described in this work. Special thanks goes to Mr. A. A. Smirnov, colleague and close collaborator since 1961, for his extraordinary help and enthusiasm in conducting investigations, results of which are presented in this book. I thank professors Yu. N. Murzenko and S. N. Klepikov and doctor V. V. Revenko for the many hours of their time they have spent on review of this work and their many valuable comments. I also express my deep gratitude to professors P. I. Yakovlev and V. N. Golubkov for encouraging me to write this book. I extend my deepest gratitude to Dr. G. P. Tsinker, President of Tsinker & Associates, Inc., Niagara Falls, Ontario, a good friend for several decades, and a strong believer in the usefulness of my work. Dr. Tsinker was instrumental in the publishing of this book; he translated the original manuscript of this book from Russian into English and committed hundreds of hours of his time for reviewing, updating and editing the manuscript. I further extend my deep gratitude to Mr. D. G. Protulipac, of Acres International Ltd., Canada, for his searching editorial work on this book, and to Ms. L. Dunn, who typed the manuscript.

George E. Lazebnik

Introduction

The economy and reliability of any kind of land-founded structure depend on a correct assessment of the stresses that are generated at the soil/structure interface (contact stresses) or the stresses within the soil mass at some distance from the underside of the structure's foundation that result from loads imposed upon the founding soil by the structure. The value and distribution of the stresses along the foundation area and in the soil mass depend heavily on the size, geometry, and stiffness (strength) of the structure itself, as well as on the parameters of the foundation soils.

Unfortunately, because very often the ground cannot be described accurately enough in terms of every possible soil/structure interaction scenario, none of the presently used analytical solutions for analyzing the soil/structure interaction phenomenon can produce accurate solutions for the stresses and their distribution at the soil/structure interface and in the soil mass. First and foremost, because the natural soil is nonuniform, its properties cannot be determined with great accuracy during the standard site exploratory procedures. Variations in natural soil properties as well as modifications of these properties in the process of construction activities and effects of the time-dependent factors make the foundation design, to some extent, hypothetical, subject to potential surprises encountered during construction.

To avoid risk, most geotechnical engineers prefer to be on the conservative side in their approach to foundation design. This has been previously described by Terzaghi (1925), Peck (1969), and, most recently, Goldstein (1982). An attempt to determine the stresses at the foundation and below by using different foundation design theories can result in up to a 20% difference in stress value (Evdokimov 1962). It should be pointed out that, numerically, the accuracy and therefore the effectiveness of the final element method, as applied to foundation design, depends greatly not only on a sound foundation mathematical model but also on the accuracy of the input data, including soil properties and soil distribution along the foundation area or contact.

The most reliable source of information for soil properties is usually obtained from field observations of existing structures fitted with reliable instruments, e.g., pressure cells, inclinometers, and other accessories used to monitor structural behavior, and to some extent from large-scale tests. Sometimes field instrumentation is used for purely scientific investigation of soil/structure interaction. However, in most practical cases it is used for the long-term monitoring of the behavior of some important structures such as nuclear power stations, large dams, important buildings, etc., constructed on compressible or difficult soils. Instrumentation allows the engineer to constantly "keep an eye" on a structure and, in the case of its unexpected behavior, such as excessive settlement, tilt, foundation overstressing, etc., to apply the appropriate remedial measures to cure results of defective design or defective construction methods. Field monitoring of structural performance allows for more economical and safer structures and more efficient execution of construction work.

Field observation of contact stresses at the soil/structure interface and in the soil mass is of particular importance. The success of these observations, particularly the long-term observations, depends on stable and trouble-free performance of the instruments used for measurement of contact stresses and stresses within the soil mass. We point out that the methods and some of the instruments currently used to measure static and dynamic stresses in foundation soils are far from ideal. Despite the great deal of progress achieved in this area in the past few decades, there are still many questions that remain to be answered, e.g., influence of size and stiffness of the instruments on reliability of their performance, the way the instruments are installed, and stress measurement methods.

Instrument size and stiffness is of particular importance because the phenomenon of soil arching around the instrument may to a greater or lesser degree affect the results obtained from soil pressure measuring cells. Also, as pointed out by Peck (1988), ". . . highly sophisticated, fully automated installations for obtaining and presenting data, sometimes held in high favor by those intrigued with gadgetry, may fail to serve a useful purpose because the simple visual observations of what may be affecting the readings are overlooked." In this sense it must be indicated that the most reliable and dependable instrument is the simplest one; it is likely to have the best chance of surviving the long period of required operation.

Before installation, soil pressure measuring instruments are calibrated. This is a very important phase of the instrumentation program. It must be carried out with a full understanding of the conditions of instrument performance during all construction and operational stages and of the loads that will act upon the instrument in the process. Last but not least, instrument design must also introduce insignificant or negligible alteration to the structure and the founding soils. In brief, geotechnical instrumentation, if properly designed, calibrated, and installed, can benefit the geotechnical aspect of the project by ensuring safety of construction, minimizing construction costs, and controlling construction procedures, and can also provide useful quantative data that is often used in advancing the science and practice of geotechnical engineering.

This book includes chapters that describe instruments used for soil pressure measurement, calibration of soil pressure cells, and methods and techniques employed in the field and in large-scale laboratory experiments, with plenty of examples of such experiments. It also provides information on research results on the investigation of the lateral soil pressure exerted upon retaining structures and the soil pressure "at rest." Because geotechnical design and construction will always be subject to uncertainties, instrumentation in general, and instruments for measurements of soil pressure in particular, will continue to be an important item of future development and improvement. I believe that in this sense this work is a useful contribution to the profession; proper use of the experimental data presented in this work can benefit the engineering judgment and can enhance calibration of mathematical models that aid in describing the phenomenon of soil/structure interaction.

1

Soil Pressure Measurement Methods: Measurement Specifics and Instruments

1.1 Direct and Indirect Methods

Normally, stresses in soils can be measured by means of indirect or direct methods. Typically, the indirect methods involve recalculation or conversion from some other measured "indirect" parameter, for example, structure deflection, to the soil pressure against structure, i.e., strain into stress. One example is indirect determination of the lateral soil pressure exerted upon a sheetpile wall. In this case the flexural strains are carefully measured through the whole height of the wall at as many points as possible. Then the distribution of the soil pressure acting upon the wall is found by the numerical differentiation of the deflection or bending moment diagrams by taking, respectively, derivatives of the fourth or second order. Practical examples of this method are given in Tschebotarioff, (1948, 1973) and Lazebnik (1961c).

The value of the soil pressure at a given level along the height of the retaining wall can also be found indirectly from the friction resistance of tapes placed on the rear side of the wall (Tschebotarioff 1948, 1973; Ohde 1938). The pressure in the soil mass can be determined similarly (Tschebotarioff 1958).

Chernyavski et al. (1987) proposed indirect methods of determination of a soil pressure diagram based on changes in the soil's modulus of deformation that occur as a result of nonuniform deformation of the footing. In this method it is assumed that the foundation soil is not uniformly compressed because of nonuniformity of the pressures transmitted to the foundation soil through the footing. The basic disadvantage of indirect methods of determination of soil pressures against structures is that, in general, numerical differentiation involves significant errors. Somewhat higher accuracy can be achieved by selection of the equation that can best approximate the bending moment diagram with subsequent differentiation of this equation. Interesting practical examples of indirect determination of lateral soil pressures against sheetpile walls are provided in Kyle and Kapp (1957), Casagrande and Wilson (1953), Casagrande (1973), Hakman and Buser (1961), and many other works.

In the direct method, the soil pressure is measured by means of a special system comprised of soil pressure cells of miscellaneous design, data recording instruments, switching devices, and communication lines. Depending on their function, the earth pressure cells are usually referred to as **embedment earth pressure cells** and **contact earth pressure cells**. The former are used for measurement of soil pressure within a soil mass, and the latter are applied for measurement of the soil pressures at the interface between the structure and foundation or backfill soils. The primary reason for using this system is either to confirm design assumptions and subsequently to provide information needed for the improvement of future designs, or to collect data in a specific research-oriented project.

Direct measurement of soil pressure is a rather difficult task. The difficulties are usually attributed to soil nonuniformity resulting from large variations in the soil's mechanical properties. When a pressure cell is located in an area of such nonuniformity or discontinuity, then its readings depend not so much on the external loads as on the state of the adjacent soils. Sometimes properties of the soil are not constant even for a uniform soil and can vary substantially depending on stress and loading conditions. This is discussed in detail in Chapter 5.

1.2 Some Specifics of Soil Pressure Measurement

The pressure cell, when placed in a soil mass, is effectively a foreign inclusion into that soil. Because deformation characteristics of the cell differ significantly from those of soil, the stressed state of the natural soil is usually distorted when stuffed with stiffer pressure cells. A stiff cell causes stress concentrations around its body. Furthermore, the cell, whose deformability is higher than that of the surrounding soil, causes a decrease in stresses in the area of its installation.

Experimental and theoretical investigations carried out by Pokrovsky (1937) and other workers helped to formulate requirements for the soil pressure cells placed within a soil mass. Not surprisingly, these studies indicated that the best results in soil pressure measurements can be achieved if the deformation modulus of the cell (E_{epc}) and that of the soil (E_s) are equal and the size of the soil pressure measurement instrument is as small as practical.

Hast (1945) and Loh (1952) suggested that the relative error in soil pressure measurement $\Delta_p = [(p_m - p)/p] \times 100\%$ (here p = stress level in soil mass in absence of load cell and p_m = stress level recorded by cell) can be expressed as a function of E_{epc} and E_s, the soil's Poisson ratio (μ_s) and the cell's depth (h) and radius (r_m) as follows.

$$\Delta_p = \frac{E_{epc}/E_s - 1}{1 + \frac{\pi r_m}{h}\left(1 - \mu_s^2\right)\frac{E_{epc}}{E_s}} \tag{1-1}$$

Baranov (1982) proposed the following slightly modified version of equation (1-1):

$$\Delta_p = \frac{1 - E_s/E_{epc}}{\frac{\pi r_m\left(1 - \mu_s^2\right)}{2h} + \frac{E_s}{E_{eps}}} \tag{1-2}$$

Note that equation (1-2) is applicable when the diameter of the pressure-bearing surface of the cell (d) is equal to its overall diameter (d_1).

Kheifitz (1973) proposed the following relationship for Δ_p:

$$\Delta_p = \frac{h}{2r_m}\frac{E_{epc}/E_s - 1}{(0.7 \text{ to } 2.0)\frac{E_{epc}}{E_s} + \frac{h}{2r_m}} \tag{1-3}$$

Comparison of the results obtained from all three formulas shows that the smallest value of Δ_p is obtained from equation (1-1), The values of Δ_p obtained from equations (1-2) and (1-3) are practically equal when the ratio E_{epc}/E_s is equal to 0.7. When this ratio is assumed to be equal to 2.0, then the calculated value of Δ_p decreases and approaches the value obtained from equation (1-1). In all three equations when $E_{epc} = E_s$, $\Delta_p = 0$; that is, no pressure concentration in the soil mass exists. This, however, is not a practical condition; it could be created only by an ideal soil pressure cell whose stiffness is the same as that of the soil. At the present time, no such cells exist.

As is generally known, deformation properties even of the same soil vary in a wide range when exposed to compressive loads. Hence, if an earth pressure cell is fabricated with a modulus of elasticity (E_{epc}) that is equal to the soil deformation modulus (E_s) at the initial stage of soil loading, then the value of E_{epc} will substantially differ from the new values of E_s with changes in load magnitude. This will result in substantial and noncontrollable soil pressure measurement errors. This is illustrated by the calibration data of two rather flexible membrane-type soil pressure cells given in Figure 1-1; the thickness of the membranes in the cells, δ_m, is respectively equal to 0.75 and 1.0 mm.

These cells are manufactured from steel and have pressure-bearing membranes of the same diameter (d_m = 50 mm) but different thicknesses. The cells have been calibrated with compressed air and by loads that have been delivered to

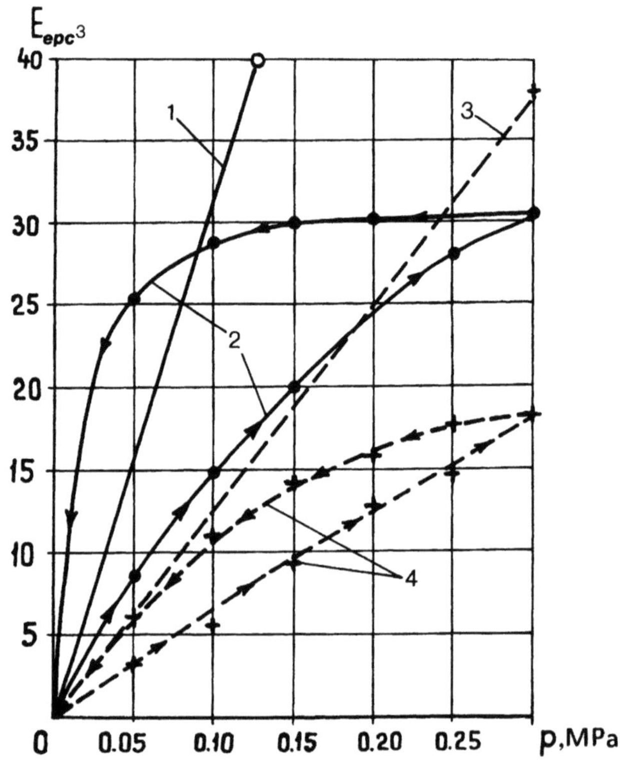

Figure 1-1. Results of calibration of membrane-type soil pressure cells (d_m = 50 mm). 1 and 3. Calibration by compressed air; thicknesses of membrane are 0.75 and 1.0 mm respectively; 2 and 4. loading and unloading of cell via noncompacted medium sand.

the cell via noncompacted medium sand. As is seen in Figure 1-1, a linear relationship exists between the readings taken from both cells and the actual pressures applied when the cells are loaded by air pressure. Drastically different results are obtained when the same cells are loaded/unloaded via noncompacted medium sand. In this case the linear relationship between cell loading and actual readings does not exist, and at a pressure p = 0.3 MPa, the value of relative error (Δp) of the earth pressure cell with membrane thickness δ_m = 1.0 mm has reached 100%, with δ_m = 0.75 mm the error has reached 200%.

Upon cell unloading, the hysteresis, i.e., the separation between loading and unloading calibration curves, is more pronounced for the cell with thinner membranes. For example, unloading of the pressure cell having δ_m = 0.75 mm from 0.3 to 0.1 MPa produced a relatively small decrease in instrument reading. On the other hand, the earth pressure cell with a higher rigidity (δ_m = 1 mm) showed a larger decrease in pressure under the same conditions. In other words, the error in cell performance decreases when the stiffness of the earth pressure-bearing membrane increases. This conclusion is supported by investigations carried out by Peattie and Sparrow (1954) and Baranov (1959). It should be noted, however, that thickening of the pressure-bearing membrane under certain conditions may result in a considerable decrease in sensitivity of the measuring system.

The calculations show that the cell's aspect ratio (height-to-diameter or radius ratio) influences the value of the error Δ_p. For example, two instruments with the same diameter of soil pressure-bearing membrane, possessing identical deformation characteristics and placed in the same soil, may (and generally do) react differently to the effects of an external load. This may be due mainly to higher stress concentrations at the thicker cell. The latter is attributed to different compressibility characteristics of both soil and cell. As illustrated in Figure 1-2, under the pressure from an external load (q) the original soil level (c-c) drops below the upper surface of the pressure cell; this results in a pressure increase upon the instrument's pressure-bearing surfaces, where $p_1 > p$.

It should be noted that attempts to develop a soil pressure cell with characteristics similar to those of soil have been initiated (Krivorotov 1971). However, to diminish the effects of stiffness noncompatibility between the pressure cell and the soil, the majority of investigators prefer to use pressure cells that are as thin and as rigid as practical; this can being the potential discrepancies between p and p_1 (Figure 1-2) to an acceptable level. The calculated results given by equations (1-1) through (1-3) show that with a high d/h ratio (thin pressure cell)

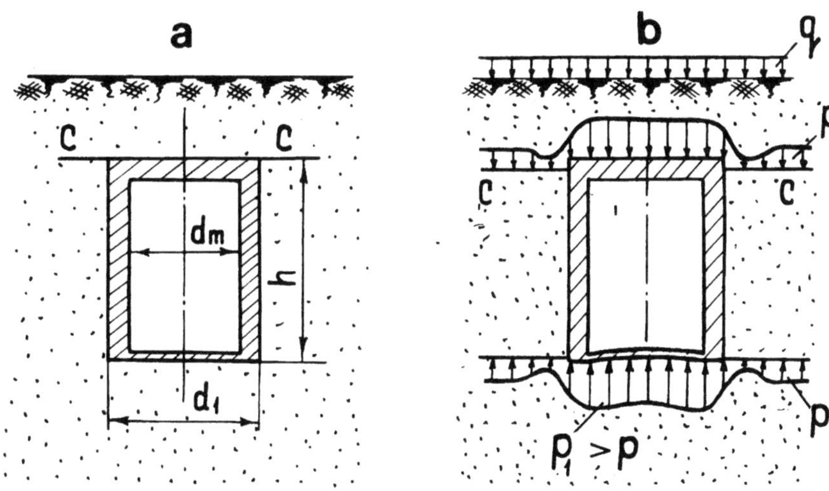

Figure 1-2. Soil pressure concentration effect at pressure-bearing diaphragms of pressure cell due to difference in compressibility characteristics of both soil and pressure cell. (a) Load is not applied; (b) after load application.

and with the same E_{epc}/E_s ratio (rigid pressure cell), the calculated error Δ_p tends to be rather small. For more information on the effects of cell aspect ratio and soil/cell stiffness ratio, the reader is referred to work by Selig (1964).

1.3 Hydraulic Multiplicator

The hydraulic multiplicator is a device that is used to increase the **sensitivity** of a soil pressure cell. The concept of this device is depicted in Figure 1-3a. Note that here and elsewhere in this work the term *sensitivity* refers to the amount of output response that an instrument or transducer produces when an input quantity is applied to it. High sensitivity does not necessarily imply high **accuracy** (degree of correctness) or **precision**. The accuracy is usually defined as the closeness of approach of a measurement to the true value of the quantity measured, and the precision is considered to be the closeness of approach of each of a number of similar measurements to the arithmetic mean.

The performance of the hydraulic multiplicator is based on properties of communicating vessels of different diameters (e.g., d and D) filled with practically incompressible liquids (the hydraulic jack effect). An example is shown in Figure 1-3a. Assume that both the large- and smaller-diameter pistons are very rigid.

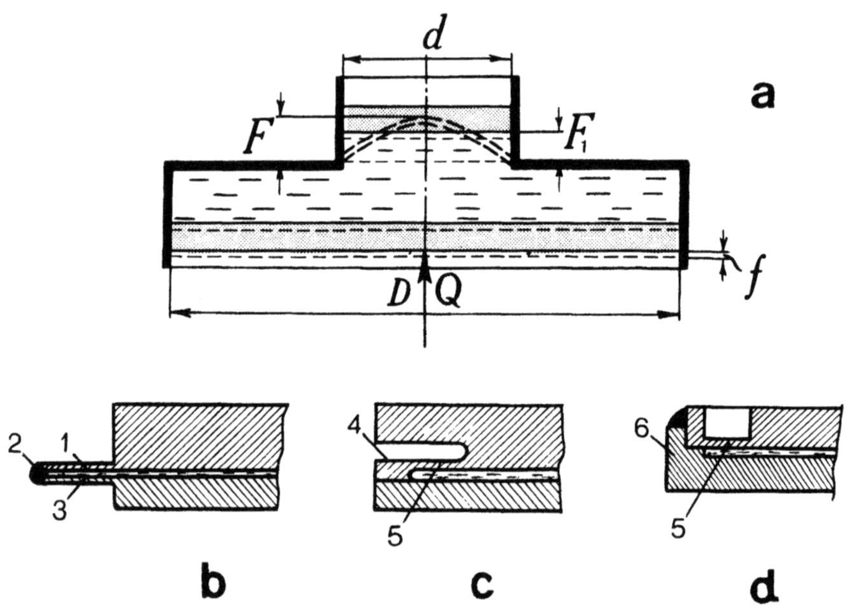

Figure 1-3. Basic concept of hydraulic multiplicator. (a) Joint between soil pressure-bearing membrane; (b) in Carlson's pressure cell; (c) in U.S. Army Engineer Waterway Experimental Station pressure cell; (d) in Baranov's pressure cell.

When the larger piston is moved a distance f by an action of force Q, then the smaller piston will be displaced a distance F_1 by action of the incompressible liquid. The relationship between f and F_1 is expressed as $F_1/f = D^2/d^2$. In practical designs the smaller piston is replaced by a membrane that can deflect slightly. Hence, with the same value of the large piston travel (f), deflection of the membrane (F) typically exceeds the movement of the very rigid small piston (F_1) by a factor approximately equal to 3. Hence, the relationship between deflections F and f can be expressed as follows.

$$F \approx 3f\frac{D^2}{d^2} \tag{1-4}$$

The design of soil pressure cells with hydraulic multiplicators involves a number of technological difficulties, including the requirement for a high precision in manufacturing of the cell's components, reliable sealing of the liquid-filled cavity, and absence of bubbles in the liquid. To reduce errors due to liquid thermal expansion, the layer of liquid must be as thin as practical. The thickness of the pressure-bearing membrane and its connection to the cell body are important because they may affect the accuracy of the soil pressure measurements.

In earlier designs of soil pressure cells the pressure-bearing membrane was made in the form of a thin plate welded directly to the cell body. In more recent designs the relatively thick membranes are joined to the cell body as illustrated in Figures 1-3b, c, and d. These types of joints allow for the manufacture of thinner cells and thus more accurate measurements of soil pressures. In a cell designed by Carlson (the stress meter, manufactured by Carlson Instruments, Campbell, California) the soil pressure-bearing membrane is joined with the cell body via thin flanges welded to each other by a continuous circumferential weld (Figure 1-3b). In a cell developed by the U.S. Army Engineer Waterways Experimental Station (Vicksburg, Mississippi) the ring groove is cut along the cell's perimeter (Figure 1-3c). This enables the cell's soil pressure-bearing membrane to uniformly transfer the soil pressures to the liquid-filled cavity. In a recent version of pressure cell designed by Baranov (1962) for the purpose of more uniform transfer of soil pressure to the liquid, a wide groove is cut in the soil pressure-bearing membrane (Figure 1-3d); the membrane is then welded to the cell body. The advantage of this design is the possibility to substantially decrease the thickness of the cell.

In all the preceding designs the stiff soil pressure-bearing membrane has minimum deflection and therefore ensures better transfer of the soil pressure to the layer of liquid. This, in turn, increases the accuracy of the soil pressure measurement.

Several types of soil pressure cells with hydraulic multiplicators have been developed and extensively used in the former Soviet Union. One of these instruments, developed by Lazebnik (1970b), is depicted in Figures 1-4 and 1-5. Here

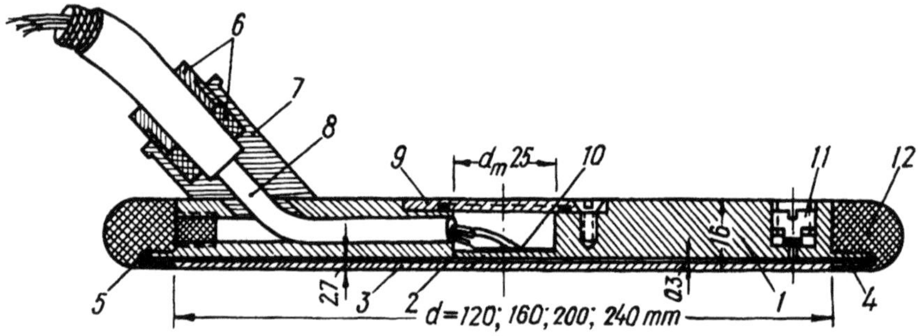

Figure 1-4. Cross section of PDGS-type soil pressure cell for determining pressure in mass of soil. 1. Body; 2. measuring membrane; 3. soil pressure-bearing membrane; 4. thin flanges; 5. circumferential weld; 6. inlet seal; 7. connecting pipe; 8. cable, 9. lid; 10. bonded resistance strain gages; 11. sealing bolt; 12. rubber ring.

and elsewhere in this book it will be referred to as the PDGS, which is the Russian abbreviation for "flat-welded soil pressure cell." This instrument is designed for measurements of stresses within a mass of soil.

Soil pressure cells of this type are fabricated in the form of two disks with soil pressure-bearing membranes of different diameters ranging from 120 to 240 mm. The depth (thickness) of this device is 16 mm. The steel body (1) is welded to the pressure-bearing membrane (3) by a circumferential weld (5). The hydraulic multiplicator is similar to that designed by Carlson (Figure 1-3b). This helps to reduce the instrument's depth. The weld and thin flanges of the instrument are

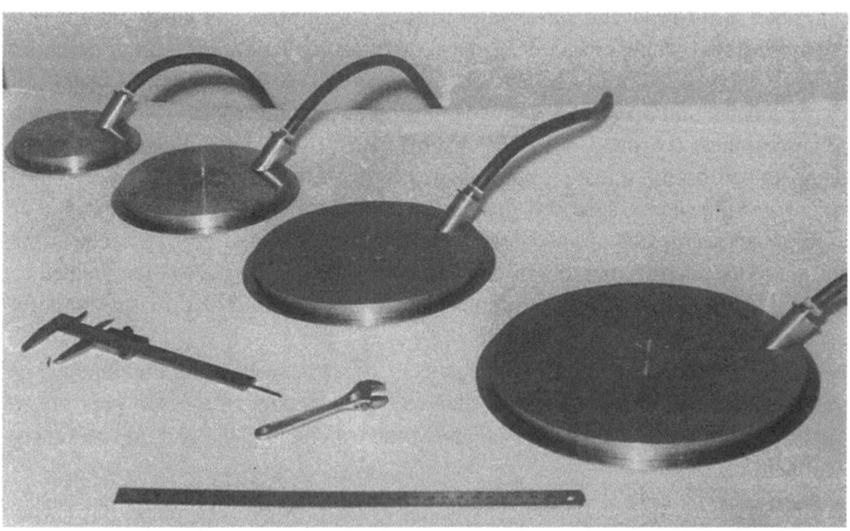

Figure 1-5. PDGS-1 soil pressure cells of several sizes. *Note*: Rubber rings are removed.

Table 1-1. Calculated Values of Modulus of Elasticity of Soil Pressure Cells (Type PDGS-1)

Diameter of Contact Surface (mm)	Deformability of Contact Surface at Pressure of 0.5 MPa (micrometers)	Modulus of Elasticity MPa
120	0.30	0.270×10^5
160	0.16	0.500×10^5
200	0.10	0.800×10^5
240	0.07	1.110×10^5

Notes: (i) in all four cells the diameter of the pressure measuring membrane is equal to 25 mm and its thickness is equal to 0.8 mm. (ii) The liquid in the hydraulic multiplicator is assumed to be noncompressible.

protected from mechanical damages by the rubber ring (12). Electrical resistance strain gages (10) are mounted on the internal surface of the measuring membrane (2) and connected either to a half-bridge or to a full external Wheatstone bridge circuit. In the latter case, the instrument's sensitivity and the stability of its operation increase. The hydraulic cavity of this cell is 0.3 mm deep and is filled with a special silicon liquid through one hole only using a special procedure. Note that in Carlson's stress meter the cavity in the multiplicator is filled with mercury.

Calculated values of the modulus of elasticity of PDGS-1 for different diameters of soil pressure-bearing membrane are given in Table 1-1. Calibration curves of the PDGS-1 soil pressure cell with contact surface 120 mm in diameter are illustrated in Figure 1-6. For calibration purposes this instrument was placed within a calibration chamber with its working surface facing upward. As is seen from Figure 1-6, because of the high stiffness and relatively small depth of the instrument, the readings are little affected by variations in soil density (deformation modulus). This applies to both loading and unloading conditions; the maximum difference between the air pressure calibration curve (1) and the curves obtained in the mass of compacted (2) and noncompacted sand (3) does not exceed, respectively, 5% and 8%. Most recently developed PDGS-type pressure cells have a total depth equal to 8 mm. In this cell the maximum difference between the air calibration curve and calibration by noncompacted sand constitutes just 6%.

Experience has indicated that for better performance of the hydraulic multiplicator the layer of liquid or mercury should be as thin as possible. For more information on the effects of grooved joints between the pressure-bearing membrane and the cell body the interested reader is referred to Dunnicliff (1988).

1.4 Orientation Effect on Cell Performance and Design

The soil pressure cells (PDGS type) described in the previous section are asymmetrical with respect to the horizontal plane. When these instruments are cali-

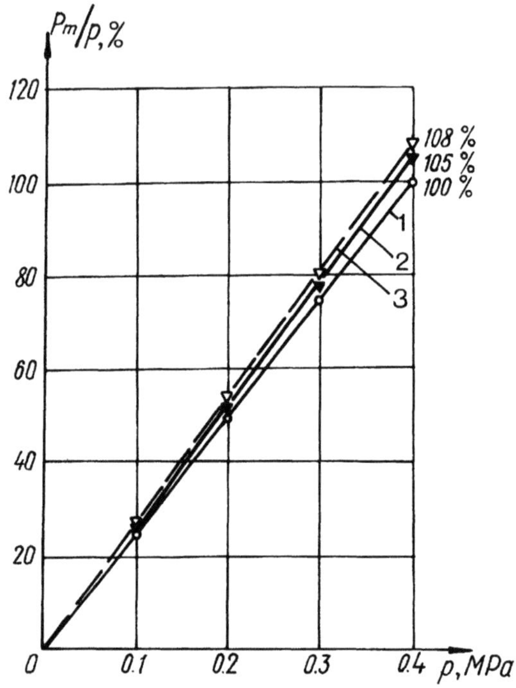

Figure 1-6. Calibration curves of PDGS-1 soil pressure cell. 1. By compressed air; 2. via compacted sand; 3. via noncompacted sand.

brated in a mass of loose soil, the "orientation effect" takes place; it manifests itself in changes in cell readings when the cell's orientation is changed in respect to the acting force. For example, when the instrument is positioned with its soil pressure-bearing surface oriented downward, the measured stresses can be in the range of 25 to 45% higher than those measured by the same cell placed with its pressure-bearing membrane oriented upward.

A practical example of the orientation effect is given in Figure 1-7. The calibration curves shown in this figure were obtained by testing the asymmetrical soil pressure cell in medium sand of different densities. These tests were carried out in calibration chamber SCC-475-05. This device is described in detail in the following section. As is seen in Figure 1-7, instrument orientation and soil density both have a substantial impact on instrument readings; when the soil is loaded by soil its readings deviate from pressure by compressed air in a range of 7% to approximately 40%.

Experiments with stiffer soil pressure cells, e.g., the PDGS-1 type described earlier, produced somewhat smaller differences in readings when calibrated by soil and by compressed air. It should be noted that in experiments with midget

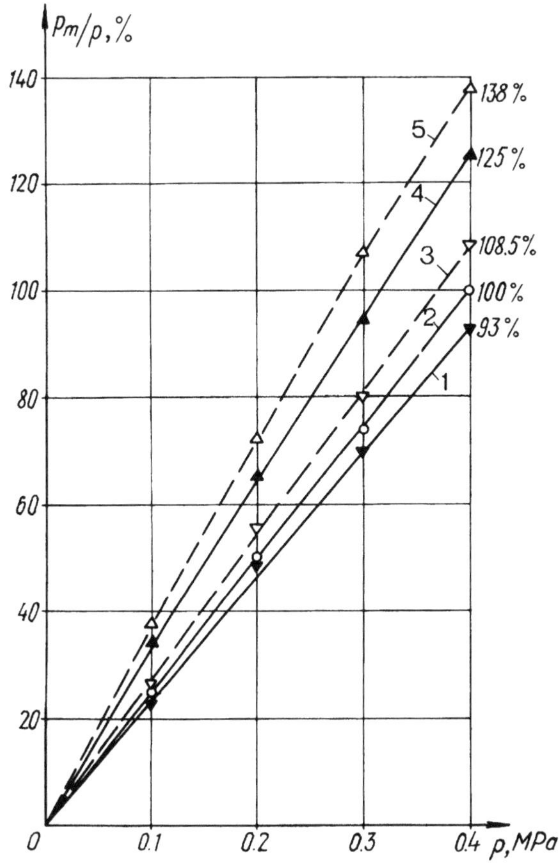

Figure 1-7. Calibration curves of soil pressure cell at different orientations within mass of soil with respect to acting force (after Uginchus and Bombchinsky 1954). 1. Loading in mass of compacted sand with soil-bearing surface oriented upward; 2. loading by air; 3. orientation the same as in 1, however, loaded in mass of noncompacted sand; 4. loading in mass of compacted sand with contact surface oriented downward; 5. orientation the same as in 4, but loaded in mass of noncompacted sand.

soil pressure cells, with the working membrane oriented downward, the deviations reached approximately 65%.

A large number of tests with pressure cells having different working membrane thicknesses, and therefore different moduli of elasticity (E_{epc}), have been performed in NIISK, Kiev. In the cells that have been tested, E_{epc} varied from 28 to 298.5 MPa. All cells were tested under similar conditions with the soil-bearing membrane oriented upward and downward. Some test results reported as $p_m/p = f(E_{epc}/E_s)$ are given in Figures 1-8 and 1-9. The cells were loaded in a special

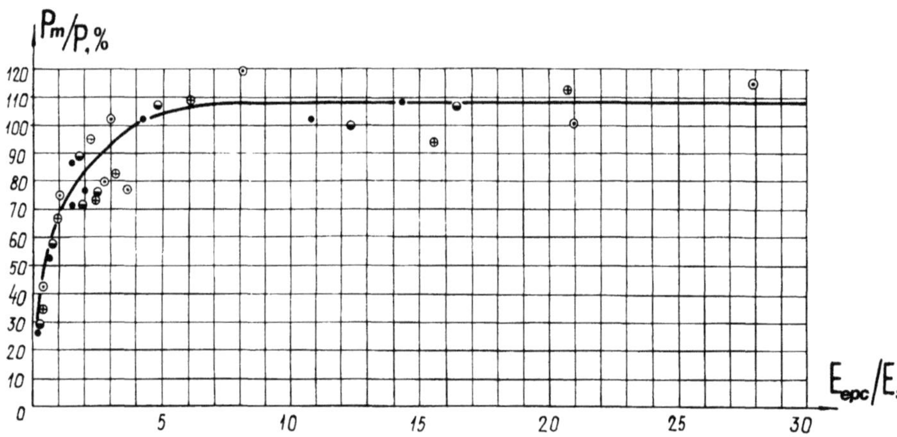

Figure 1-8. Relative value of soil pressure (p_m/p) measured by pressure cell in mass of soil as function of E_{epc}/E_s. *Note*: Soil pressure-bearing membrane is oriented upward.

calibration chamber filled with sand of various granulometric compositions and with different values of E_s. Each point in the curves presented in Figures 1-8 and 1-9 is an average from 3 to 5 identical tests.

From the curves presented in Figures 1-8 and 1-9 it is possible to conclude that the equality of the effective (p) and the measured (p_m) soil pressures ($p_m/p = 100\%$) for the particular type of soil and the membrane pressure cell can be achieved when the ratio $E_{epc}/E_s \approx 4$ and the cell is placed with its soil-bearing

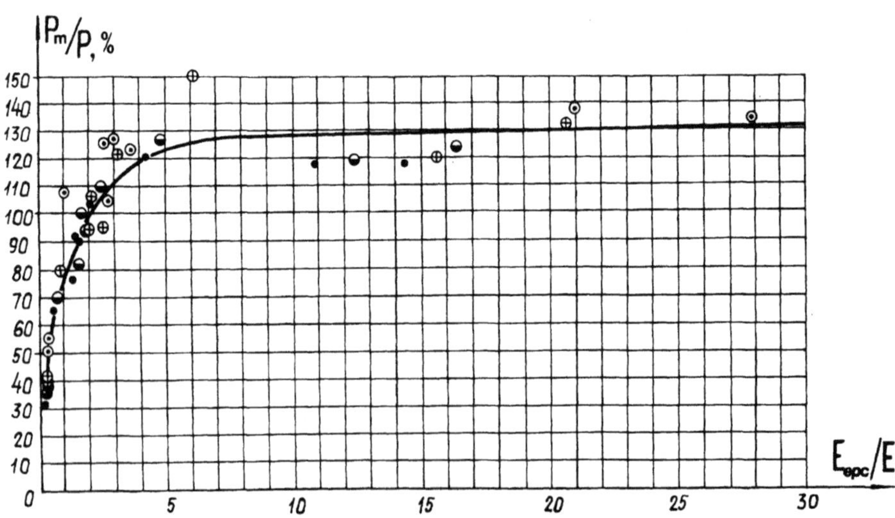

Figure 1-9. Relative value of soil pressure (p_m/p) as function of E_{epc}/E_s. *Note:* Soil pressure-bearing membrane is oriented downward.

membrane oriented up, and when $E_{epc}/E_s \approx 2$ with the membrane oriented downward. It is interesting to note that the lower branches of both curves up to $p_m/p \approx 60\%$ ($E_{epc}/E_s \simeq 0.5$) are practically identical. The maximum difference in p_m and p readings ($\approx 30\%$) exists at $E_{epc}/E_s \approx 4$ to 10 when the cell is placed with its working membrane oriented downward.

Special investigation of the stressed soil condition that is developing around the soil pressure measuring instrument have been carried out in order to better understand the nature of the orientation effect on cell readings. This has been carried out in a large calibration device illustrated in Figures 1-10 and 1-11. Special large diameter instruments were used in this investigation. Some characteristic parameters of these instruments are depicted in Table 1-2. These and the other comprehensive studies have been carried out in NIISK, Kiev and results

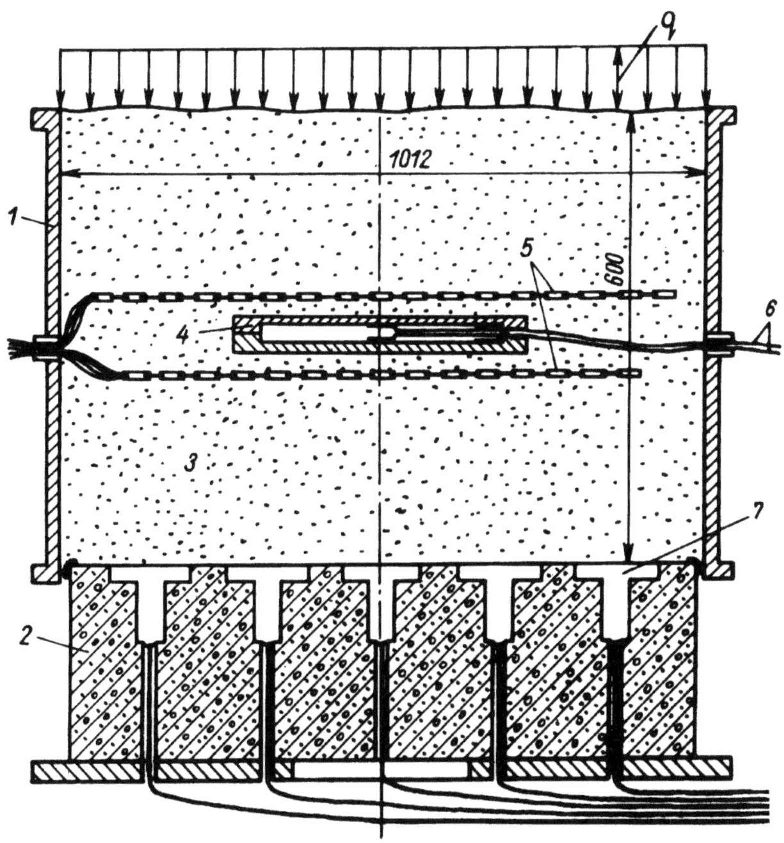

Figure 1-10. Cross-section of calibration chamber for calibration of large soil pressure measurement devices. 1. Steel tank; 2. concrete base; 3. sand; 4. embedment soil pressure cell; 5. miniature soil pressure cells; 6. cables; 7. rigid soil contact pressure cells.

Figure 1-11. Installation of large soil pressure cell, shown in Figure 1-10, in calibration chamber.

of this investigation are reported by Lazebnik (1969; 1970a and b; 1974a). The basic conclusions from this study are as follows.

- The difference in readings when the cell is placed upwards and downwards with respect to load application is more pronounced when the cell is placed in noncompacted soils.

Table 1-2. Selected Parameters of Sensors with Large Pressure Bearing Membrane

Designation of sensor	Designation of a half of membrane	Membrane thickness, mm, ±0.05	Stress developed in membrane at p = 0.1 MPa, MPa	Deflection of membrane at p = 0.1 MPa, in micrometers	Total modulus of elasticity of sensor E_{epc} MPa and schematic cross section (dimensions in mm)
A	I A	10	30	130	28.2 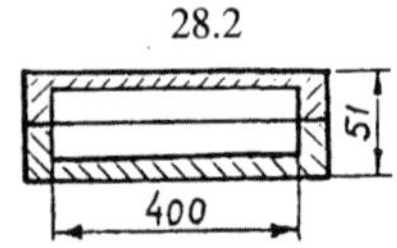
	II A	13	16	50.6	
B	I B	18.7	8.6	20	212 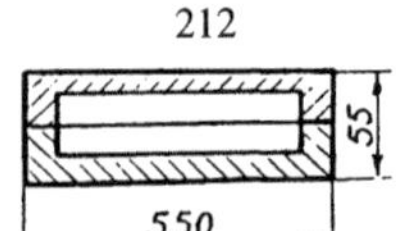
	II B	28	3.8	6	

- Stress concentration is formed under sufficiently rigid instruments; when the lower membrane of the pressure cell is not rigid, then stress concentration may not be developed at all. This phenomenon has been observed during tests with large cells (Figure 1-11), with the heaving thickness of the soil pressure-bearing membrane equal to 10 mm.
- The adverse effects of cell orientation can be overcome by use of a symmetrical type of soil pressure measurement device. The concept of such an instrument (Figure 1-12) was proposed by Lazebnik (1969).

Cell placement effects (e.g., installation of cells upon not perfectly leveled soil surfaces) may also be the reason for different readings recorded by the opposite active faces of the flat cell. Binnie et al. (1967) comment that the initial difference in cell readings can be caused by nonplanar leveling of the soil surfaces upon

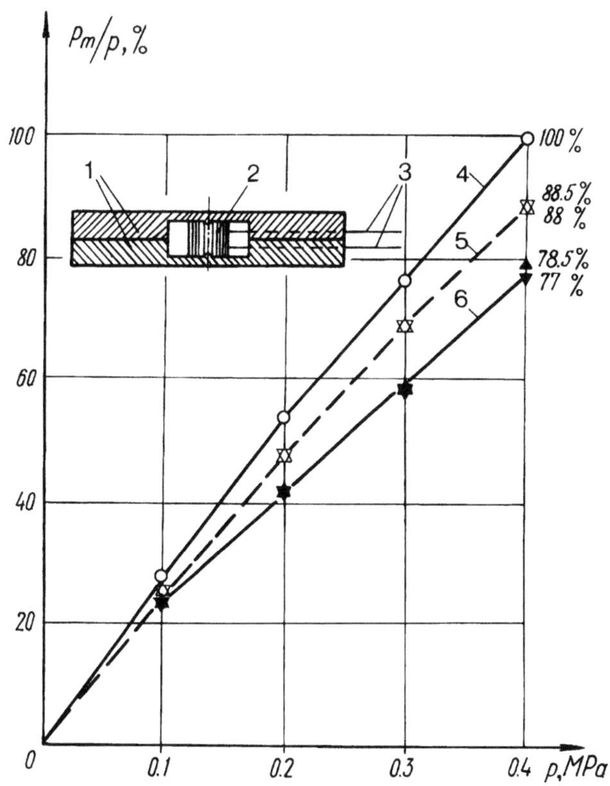

Figure 1-12. Nonrigid soil pressure cell; cross section and calibration curves (Lazebnik 1969). 1. Identical halves; 2. transducer; 3. wires; 4. calibration curve by compressed air; 5. calibration curve in mass of noncompacted sand; 6. calibration curve in mass of compacted sand.

which these cells are placed, whereas a part of these differences may be due to stress variations in the fill caused by the cells themselves. Binnie et al. also report on substantial underreadings of the instrument and differences in readings between both faces when the same flat cells were installed as embedment cells in gravel.

Uff (1970) demonstrates the high sensitivity of a flat cell with two active faces to uniformity of bedding by installing it as a contact cell. Thomas and Ward (1969) report that flat cells with two active faces and vibrating wire type transducers installed in clay register small differences between both faces. The latter investigators and Kennard et al. (1967) observe that the same cells placed vertically in the clay core of a dam, as might be expected, produce good agreement in readings on both sides of the cell. Wilson (1984) attributes the difference in double active face cell readings to difficulties in leveling the bedding for the cell with sufficient accuracy; in his judgment this may deform the cell and cause additional errors in cell readings.

As noted earlier, Lazebnik (1969) proposed the concept of a soil pressure cell symmetrical about its middle plane (Figure 1-12). This instrument consists of two identical halves. Each of these halves represents a disk with a soil-bearing membrane located in the middle part of the disk. The pressure measuring transducer is provided between the membranes; the variation in electrical parameters of the transducer are proportional to the value of deformation of the membranes between which it is installed. A total of five instruments of this type have been manufactured and tested in calibration chamber; typical calibration curves are given in Figure 1-12. These curves are based on the calibration results of a pressure cell having soil pressure-bearing membranes 2.6 mm thick.

Each point on the calibration curves was obtained from 4 to 12 tests. Despite the fact that the test cells are symmetrical by design, they were loaded after being placed within the calibration chamber both upward and downward. In these tests readings of a cell loaded by air are taken as 100%. As is seen from Figure 1-12, the calibration curves of the symmetrical soil pressure cells are practically identical in soils having close values of the modulus of deformation, i.e., the test points are superimposed. A very small scatter of test points obtained in calibration curve 6 ($p = 0.4$ MPa) is explained by the impossibility of creating completely identical conditions in different tests.

The basic concept of the symmetrical cell, as described above, was used for developing several types of instruments for measurement of the pressure in a soil mass. Typical cross sections of two modifications of these cells are depicted in Figure 1-13. The superior performance of these cells in comparison with other instruments is due to the use of stiff soil pressure-bearing membranes and the hydraulic multiplicators. These cells proved to be reliable and demonstrated an outstanding longevity and performance stability during long-term monitoring of the stress condition in a soil mass.

The instrument shown in Figure 1-13a is recommended for measurements of static pressures in a soil mass. It consists of two nonsymmetrical cells (1) placed

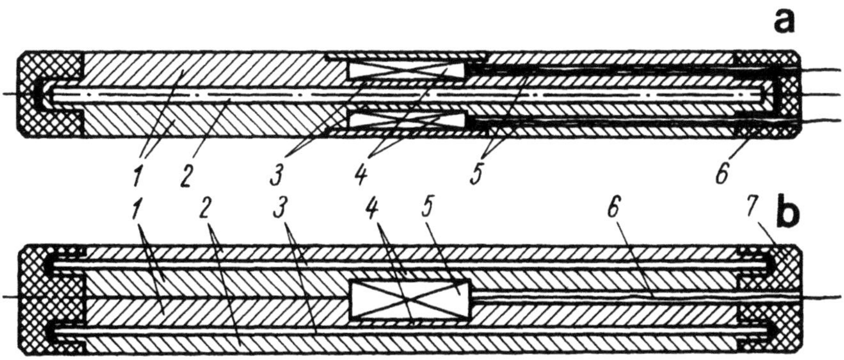

Figure 1-13. Rigid symmetrical soil pressure cells relatively free from orientation effects. Typical cross sections. (a) With one layer of liquid 1. cell body consisting of two identical halves; 2. liquid-filled cavity; 3. pressure measuring membranes; 4. transducers; 5. wires; 6. rubber ring. (b) With two layers of liquid 1. body; 2. contact surfaces; 3. liquid-filled cavities; 4. membranes; 5. transducer; 6. wire; 7. rubber ring.

back-to-back and welded together. Both halves share the same layer of liquid (2), which is designed to transfer the soil pressure from the soil contact plates (membranes) to the pressure measuring membranes (3). The soil pressures are measured by two independent transducers (4) in which bonded wire or bonded soil resistance strain gages are used. [For details on this kind of gage the reader is referred to Dunnicliff (1988); information on these gages is also found in other works referenced elsewhere in this book.] The use of two duplicate transducers substantially enhances the cell's long-term performance.

The pressure cell depicted in Figure 1-13b is comprised of two nonsymmetrical cells, each having its own hydraulic multiplicator. Both cells are welded back-to-back, and soil pressure measurement by both cells is carried out by a single transducer (5) located in the middle of the instrument. The soil contact membranes of this cell (2) are relatively thin, which contributes to the small instrument mass. This makes the instrument practical for measurement of both static and dynamic soil pressures.

Typical calibration curves for the soil pressure cell depicted in Figure 1-13a are presented in Figure 1-14. From these curves it is seen that the maximum difference between the calibration curve obtained from tests by the compressed air (1) and those obtained from tests in compacted (2) and noncompacted soils (3) does not exceed 8%, and the cell orientation effect barely exists. Similar features are achieved in miscellaneous modifications of symmetrical pressure cells developed by Lazebnik (1970b). A general view of different modifications of symmetrical pressure cells is shown in Figure 1-15.

Green (1986) suggested that the best performance of soil pressure cells having hydraulic multiplicators can be achieved when the cell is created with two back-to-back grooved hydraulic multiplicators, with stiff membranes and each with

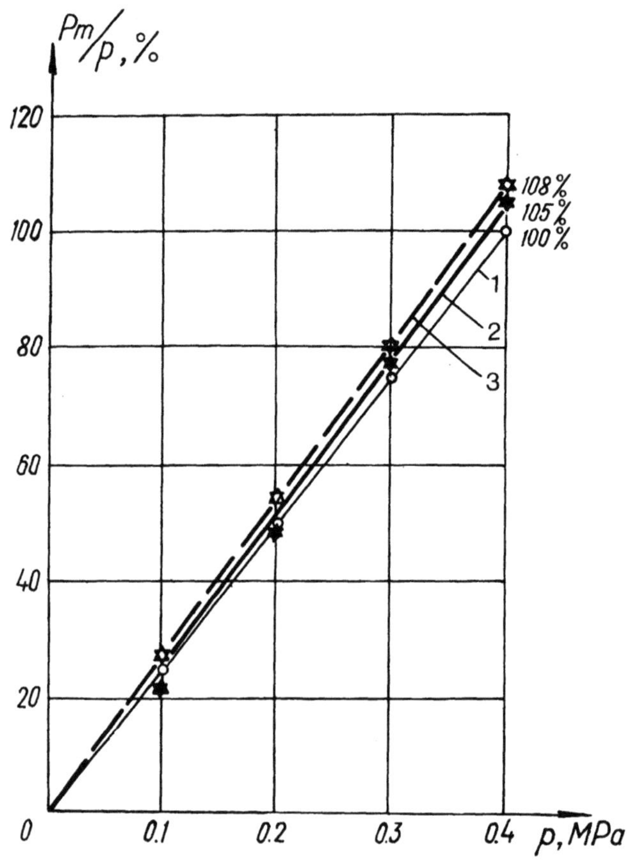

Figure 1-14. Calibration curves of soil pressure cell depicted in Figure 1-13a. 1. Loading by compressed air; 2. loading in mass of compacted sand; 3. loading in mass of noncompacted sand.

its own transducer; the cell would therefore provide two independent measurements and would have minimum sensitivity to nonuniformly leveled soil bedding.

In concluding this section, it should be noted that long-term monitoring of changes in soil pressures is best achieved by using electroacoustic (vibrating wire) transducers. This type of transducer, however, is not suitable for measuring dynamic stresses in the soil mass and at the soil/structure interface. A more detailed discussion on both strain gage and vibrating wire transducers is given elsewhere in this book. For an in-depth discussion on vibrating wire systems see Thomas (1966) and Dreyer (1977). For more information on the effects of cell orientation in soil on soil pressure readings the reader is referred to DiBiagio (1977), ISRM (1981), Weiler and Kulhawy (1982), and Wilson (1984).

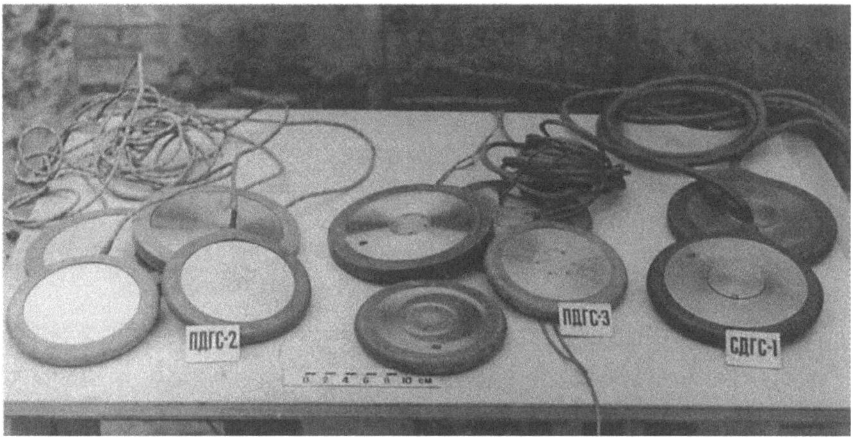

Figure 1-15. Symmetrical soil pressure cells of miscellaneous modifications.

1.5 Cell for Measuring Both Static and Dynamic Soil Pressures

Many structures that are constructed upon foundation soils are exposed to both static and dynamic loads. These loads are usually measured separately by dedicated instruments. As pointed out earlier, the instruments with electrical resistance strain gage transducers or vibrating wire type soil pressure measuring systems are typically used in cells designed for measuring the static pressure in a soil mass. It has been also noted that instruments with vibrating wire transducers are particularly preferred where long-term observations of static soil pressures are required.

Lazebnik (1974a and 1980) developed an instrument that is able to take measurements of both static and dynamic pressures in a mass of soil. A cross section of this instrument is depicted in Figure 1-16. This instrument is comprised of two halves welded together having different masses; one half with the smaller mass is designed for measuring stresses produced by dynamic loads, and the other one of larger mass is designed to measure stresses attributed to static loads. A vibrating wire type transducer is installed into the lower half of the cell with the smaller mass; the latter half is provided with a bonded soil resistance strain gage system. A layer of liquid is provided between both halves of the instrument. This helps transmit the average soil pressure, taken up by the upper and lower soil pressure-bearing surfaces of the instrument, to the measuring membranes of both transducers. The space around the transducer and inlets for wires and cables are sealed with epoxy cement and a special sealing compound. The hydraulic multiplicator principle is also used in the design of this instrument to help increase the instrument's sensitivity.

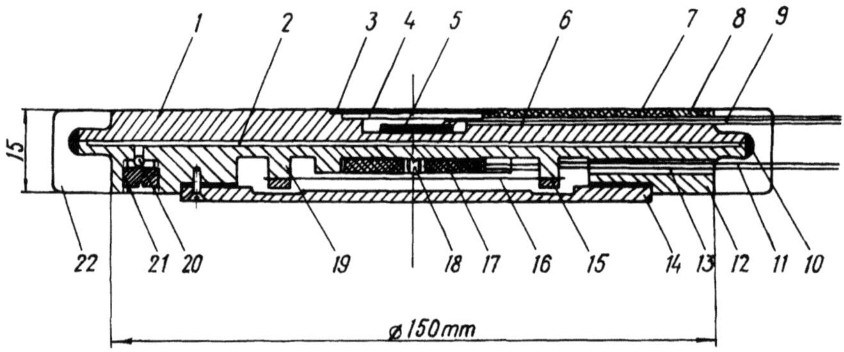

Figure 1-16. Soil pressure cell for measurement of both static and dynamic stresses in mass of soil. 1. Plate with strain gage transducer (SGT); 2. hydraulic cavity; 3,4. respectively, external and internal lids of SGT; 5, bonded foil resistance strain gage transducer; 6. groove for SGT wires; 7. sealing compound; 8. pin; 9. SGT wires; 10. circumferential weld; 11. sound wire transducer cable; 12. plate; 13. groove for cable; 14. lid; 15. sound wire fastening plate; 16. sound wire; 17. induction coil; 18. core; 19. string holding bracket; 20. steel packing ball; 21. bolt; 22. rubber ring.

Typical calibration curves of this cell loaded by air pressure and in sands with different moduli of deformation are presented in Figure 1-17. All curves are practically rectilinear. The unloading curves are not shown here because they practically coincide with the loading curves. From this it follows that this instrument is rigid enough so that its readings essentially do not depend on the variation in changes of the modulus of soil deformation which practically occur in the process of load application and removal.

Although the soil pressure cell in question is almost symmetrical about its middle plane, the different thicknesses (masses) of both halves constituting its body result in some orientation effects. The investigations have shown that the sound wire transducer (WT) does not manifest this effect. However, when the bonded soil resistance strain gage transducer (SGT) is placed in the soil mass with its lighter half oriented downward, it results in an excess of measured stresses by approximately 7.4% on average, as compared with readings taken when this same half is oriented upward. In practice, however, because the instrument is almost always placed in both a mass of soil and in a calibration chamber with its lighter half oriented upward, the orientation effect does not present a real concern. For more information on the design, construction, and performance of a diaphragm cell with two active soil pressure-bearing membranes, the reader is referred to work by Thomas and Ward (1969).

In conclusion, it should be pointed out that in the past 30–40 years several types of diaphragm and hydraulic soil pressure measurement instruments have been developed and practically implemented. These are discussed in works by Baranov et al. (1971b), Budin and Nerpin (1962), Eidelman (1964), Kheifitz

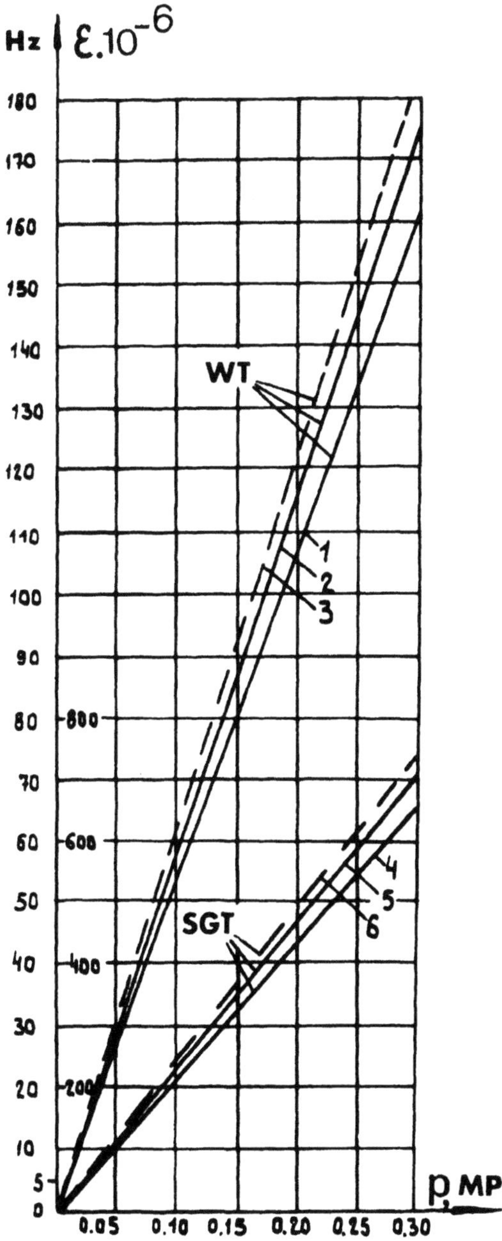

Figure 1-17. Calibration curves of the soil pressure cell type depicted in Figure 1-16. WT: Sound wire transducer; SGT: Bonded foil resistance strain gage transducer. 1 and 4. Loading by compressed air; 2 and 5. and 3 and 6. loadings, respectively, in compacted and noncompacted sand.

and Petrashen (1973), Kuznetsov and Budko (1961), Loh (1952), Peattie and Sparrow (1954), Barker and Reese (1969), Carlson (1975), Green et al. (1983), Nowack and Gartung (1983), Price and Wardle (1983), Horwath (1985), and many others. The state of the art of the subject matter is discussed in detail by Dunnicliff (1988).

2

Contact Soil Pressure Measurement Techniques

2.1 Conventional Approach and Experimental Verification

Two basic techniques are normally used for the direct measurement of contact soil pressures at the interfaces between the structure and founding soil. In one of these techniques, the contact soil pressures are measured by means of soil pressure cells installed at the interface between the soil and the underside of the structure; in this case the soil pressure-bearing membrane (contact surface) of the instrument is placed flush with the exposed-to-soil pressure face of the structure, e.g., underside of the footing, backside of the retaining wall, tip or sides of piles, etc. In the second technique, the soil pressure cells are placed in the mass of soil at some distance (usually approximately 30 to 100 mm) from the underside of the footing. Some advocates of this technique (Eidelman 1960a, b; Murzenko 1965; and others) believe that this eliminates the effects of potential soil unevenness at the interface with a structure on readouts by soil pressure measurement instruments.

In the early 1960s Lazebnik with co-workers carried out investigations to determine the merits and disadvantages of both techniques (Lazebnik 1966; Lazebnik et al. 1966). The investigations were performed in a soil pit filled with fine river sand having a natural water content between 2.6 to 2.8%; sand was placed into the pit in layers and compacted to a void ratio value of $e \approx 0.64$; the surrounding soil within the testing pit consisted of dusty, sandy loam and had approximately the same void ratio. The investigations were carried out using two stiff reinforced concrete blocks, the size and geometry of which are illustrated in Figures 2-1 through 2-3. The load on these blocks was applied by a hydraulic jack in a manner similar to that used in standard pile tests. The load bearing beam of the block loading installation was supported on four piles, two at each end of the beam. The soil pressure cells used for measuring the contact pressures had a stiffness comparable to that of concrete.

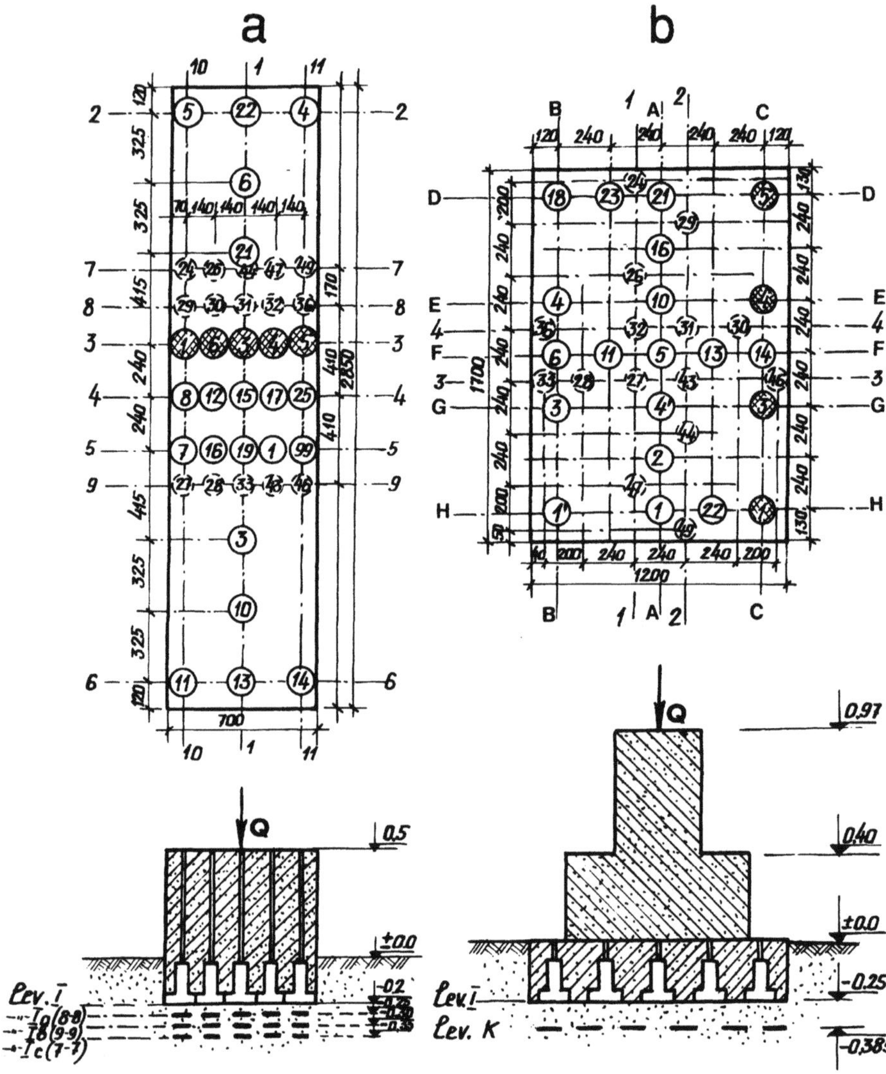

Figure 2-1. Geometry of reinforced concrete blocks: (a) and (b) plans and cross sections.

Legend:

contact pressure cells with sound wire type transducers

contact pressure cells with bonded foil resistance transducers

— pressure cells for measurement of soil pressure in mass of soil.

Figure 2-2. General view of test block depicted in Figure 2-1a.

Figure 2-3. General view of test block depicted in Figure 2-1b.

All cells had hydraulic multiplicators. Their soil pressure-bearing membranes (diameters 135 and 132 mm) were stiff. Some of the instruments had a sound wire transducer and the others had transducers based on bonded wire resistance strain gages. A detailed description of these instruments is given in Chapters 3 and 4. Soil pressure cells were then installed in the test blocks with their contact surfaces flush with the concrete surface. Each cell was tightly installed in a special steel nest embedded into the concrete.

To measure the contact soil pressures in accordance with the second technique, the soil pressure cells were placed in the soil mass at distances of 5, 10, 13.5, and 15 cm beneath the test blocks, as indicated in Figure 2-1. This was done to determine the optimum distance from the underside of the footing to instruments intended for measurements of soil contact pressures. The principles of soil pressure cells used for measurement of stresses in soil mass are discussed in Chapter 1. To avoid interference with each other, all pressure cells were placed in a staggered pattern.

The load on the test blocks was applied in increments of 100 kN, each with a mean nominal uniformly distributed pressure of approximately 0.05 MPa under each block. The mean maximum pressure under the block depicted in Figure 2-1a was brought up to 0.35 MPa; under the block depicted in Figure 2-1b it was brought up to 0.3 MPa. The unloading was carried out in increments of 0.05 and 0.1 MPa. Each of the two test blocks was loaded and unloaded twice. The test results are illustrated in Figures 2-4 through 2-6.

Comparison of the soil pressure diagrams obtained by contact pressure cells installed directly into the test blocks with those obtained by soil pressure cells installed in the soil mass underlying these blocks, particularly in the first set of readings, indicates that, despite the careful installation of pressure cells in the soil mass and meticulous preparation of the founding soil, cell readings in both methods of soil pressure measurements produced substantial discrepancies. This was most characteristic for the initial stages of load applications that resulted in pressures upon the foundation soil of up to 0.2 to 0.25 MPa. The differences are largely attributed to some imperfections left over in the local soil densities as well as in surface leveling. However, further increases in load upon the founding soil resulted in better contact between the test block and the soil, and consequently smaller discrepancies in the cells readings were obtained.

Much better results were obtained in the second set of loadings, where good performance of all soil pressure cells was registered, even when small loads were applied. Qualitatively, the geometries of the soil pressure diagrams in this case are comparable to those obtained during the first set of loadings. Some characteristic results of both sets of loadings obtained from the test block assembly depicted in Figure 2-1a are presented in Figure 2-4. These contact soil pressure diagrams were obtained by contact pressure cells installed directly in the test blocks.

The total area of the soil pressure diagrams obtained from these tests approximates pretty closely the effective loads that actually have been exerted upon the

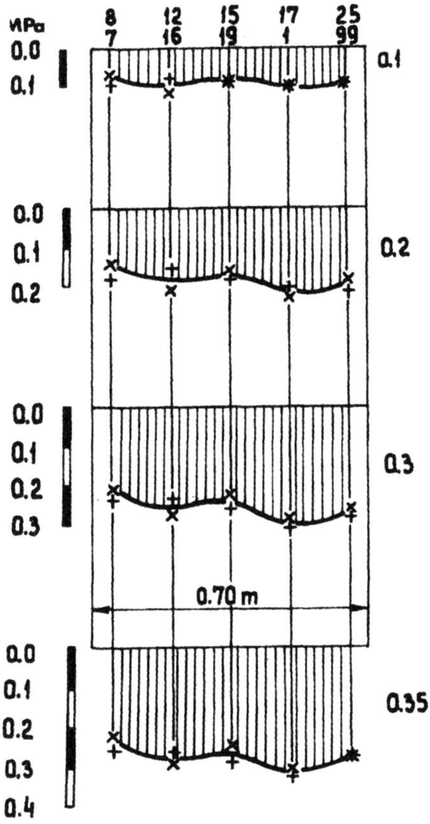

Figure 2-4. Contact soil pressure diagrams obtained by means of soil pressure cells equipped with vibrating wire type transducers (sections 4 and 5 of test block (Figure 2-1a).

test block, e.g., the difference at mean pressures of up to 0.25 MPa does not exceed 8 to 10. (Note that the measured reaction of the foundation soil happened to be somewhat smaller than the actual applied load.) As the load increases and the pressure on the soil increases up to 0.3 to 0.35 MPa, the aforementioned difference also increases and reaches approximately 14%. The imbalance of applied and measured loads is best explained by the effects of friction between the soil and areas of test blocks in contact with the soil.

From Figures 2-5 and 2-6 it is seen that the areas of the soil pressure diagrams obtained by means of pressure cells installed in the underlying mass of soil depend on the distance between these cells and the underside of the test block. From these diagrams one may conclude the following: in the case where soil pressure measuring cells are installed at 5 cm beneath the test block, the total

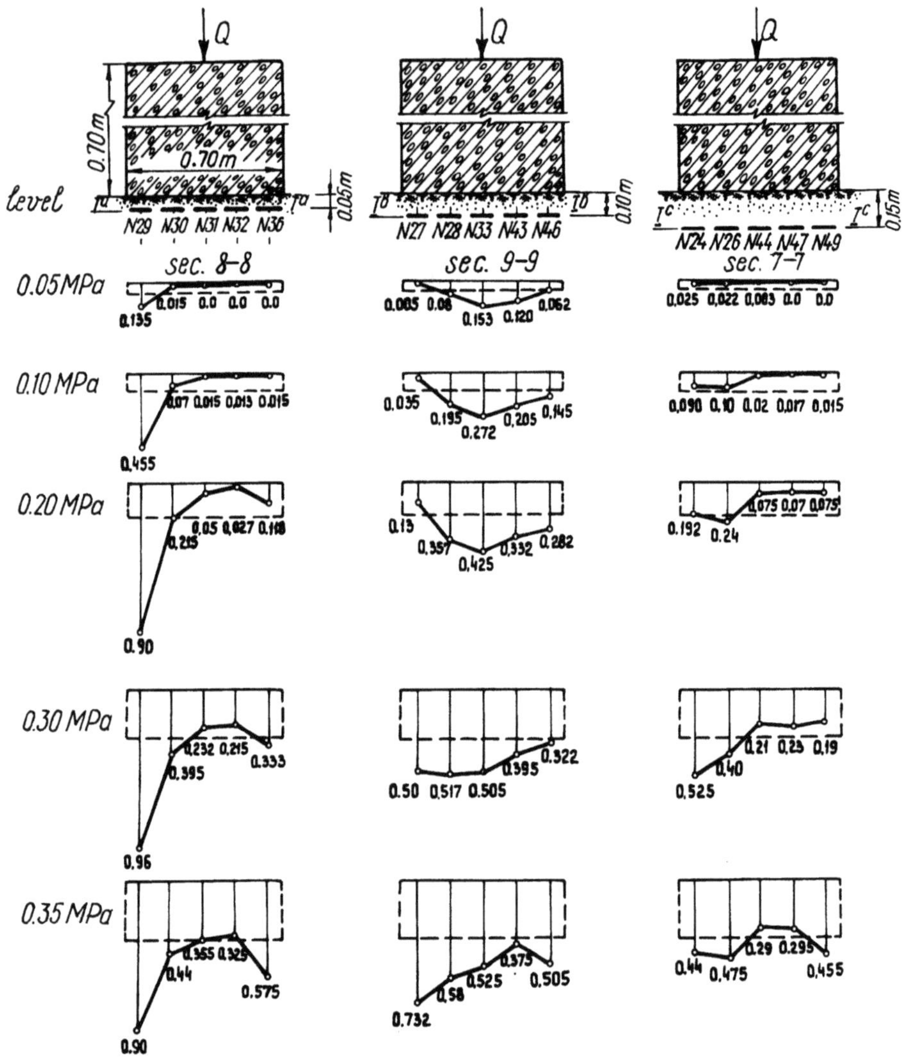

Figure 2-5. Contact soil pressure diagrams in underlying mass of foundation soil in sections 7, 8, and 9 (Figure 2-1a) in first loading cycle.

measured reaction of the soil exceeds the applied load by 17 to 20%; when the cells are placed 10 cm beneath the block, the measured reaction of the soil is approximately equal to the applied load; and when the cell placement distance is 15 cm, the measured soil reaction is smaller than that of the applied load. Similar results have been obtained from experiments involving the test block assembly depicted in Figure 2-1b.

Although the preceding phenomenon is difficult to explain, the obvious conclu-

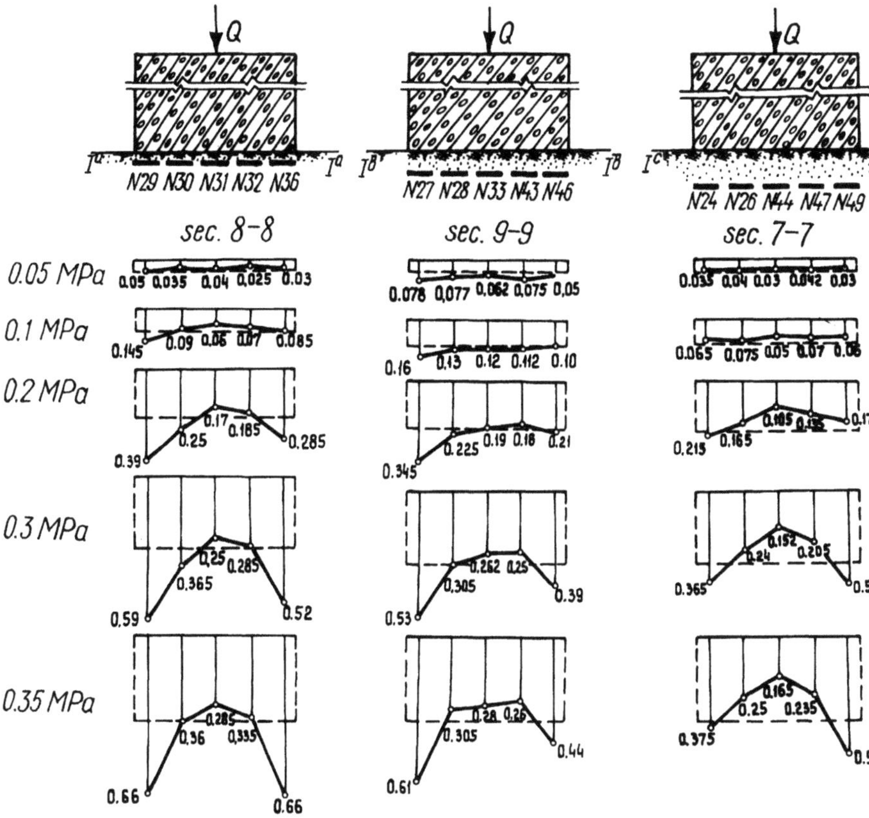

Figure 2-6. Contact soil pressure diagrams in underlying mass of soil in sections 7, 8, and 9 (Figure 2-1a) in second loading cycle.

sion is that there must be an optimal depth for placement of soil pressure measurement instruments when the second soil pressure measurement technique is employed. For this, the foundation geometry, the relative size of the soil pressure measuring cells, as well as the instrument's ability to register both vertical and horizontal components of soil pressures with adequate sensitivity must be taken into account.

In the case in question, similarly to the case of measurements of contact soil pressures by means of contact pressure cells, much more consistent results were obtained in the second set of loadings. The ordinates (dependent variable) of the contact soil pressure diagrams for equal loads, obtained by using both techniques, are approximately equal at the middle part of the test blocks. However, the contact soil pressures measured by means of cells installed in the underlying mass of soil is 1.5 to 2.0 as much as that measured by contact cells installed directly in the test blocks. A typical example is illustrated in Figure 2-7. Again,

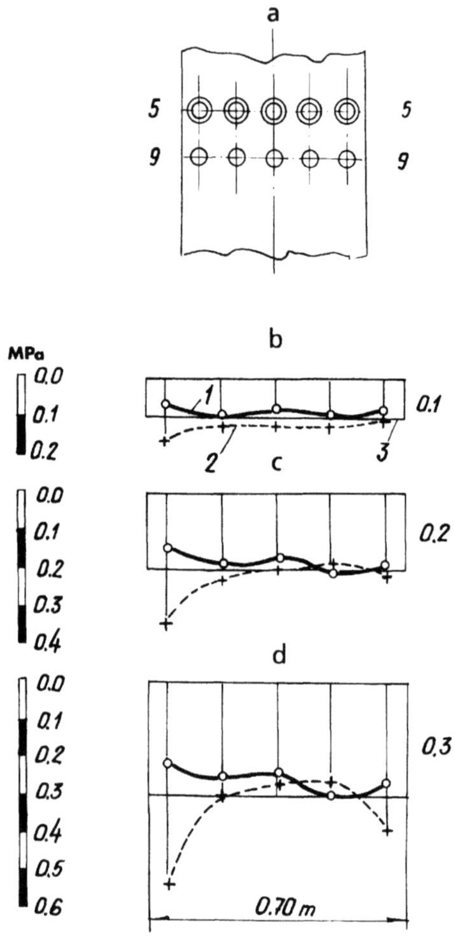

Figure 2-7. Test block depicted in Figure 2-1a; soil pressure diagrams obtained at different sections: (a) Layout of soil pressure cells; (b, c, and d) pressure diagrams due to applied loads equal to 0.1, 0.2, and 0.3 MPa respectively. 1. Plotted from readings of contact pressure cells at section 5-5; 2. plotted from readings of cells installed in mass of soil 100 mm beneath the block's underside (section 9-9); 3. applied uniformly distributed load.

this phenomenon is difficult to explain; most likely it has been caused by the effects of horizontal components of compressive stress on pressure cell readings.

2.2 Stress Deviator and Its Effects on Variation in Soil Pressure Cell Readings

The effects of the stress deviator; that is, σ_r/σ_z apparently was first discovered by Plantema (1953). Here σ_r and σ_z, respectively, are lateral and vertical stresses in the soil mass. While calibrating pressure cells in a cylindrical device, Plantema found that the slope of the calibration curves for a flat soil pressure cell at $\sigma_r/\sigma_z = 0.33$ differs from that obtained at $\sigma_r/\sigma_z = 0.5$. He also discovered that the sensitivity of the soil pressure cell increased at higher values of σ_r. It should be noted that this phenomenon is not directly discussed by Plantema (1953); however, it is clearly present in the graphical forms of his work. Kheifitz and Petrashev (1973) confirmed that lateral soil pressure may have significant effects on soil pressure cell readings. They found that variations in $\sigma_r/\sigma_z = (0.48$ to $0.4)$ could result in differences in cell readings in a range of 10 to 50%.

In the late 1960s and early 1970s Lazebnik and his co-workers carried out an in-depth investigation of this phenomenon (NIISK, Kiev, unpublished reports). The following is a brief description of these investigations and the results that have been obtained. The investigation was carried out in a special calibration chamber having very rigid walls and bottom. This helped to strictly control the ratio σ_r/σ_z during these investigations. The chamber was designed to sustain a maximum pressure of 0.5 MPa. A detailed description of this device is given in Chapter 3. Dry medium sand was used in these investigations. The lateral, σ_r and vertical, σ_z, pressures were measured by means of miniature pressure cells installed at three levels (65, 130, and 195 mm) from the bottom of the device; three cells at each level were placed at a distance of 40 mm from the chamber's walls. To eliminate the possibility of interference with each other, all instruments were installed in a staggered pattern.

The ratio of lateral (radial)-to-vertical stresses (σ_r/σ_z) was calculated as an averaged value of readings from the aforementioned cells, arranged in the mass of soil both vertically and horizontally. During all these tests the values of σ_r/σ_z were found to vary in a range of 0.53 to 0.57.

In the first series of tests the soil pressure cells were loaded in a mass of soil without provision for its lateral expansion. The readings of the instruments recorded in this series of tests have been conditionally taken as 100%. The second series of tests were carried out with provision for some lateral expansion of soil. For this a sheet of compressible material (porolon) 5 mm thick was pasted on the inner walls of the calibration device before placing the soil into it. Under such conditions the maximum relative radial deformation of soil, $\varepsilon = \dfrac{\Delta R_{ca}}{R_{ca}} \approx 2 \times 10^{-2}$. Here, R_{ca} is the internal radium of the calibration chamber and ΔR_{ca} is the increase in R_{ca} due to compression of the chamber's lining. With this expansion, the radial-to-vertical stress ratio in soil $\sigma_r/\sigma_z = (0.44$ to $0.38)$ was markedly smaller than that obtained in the first series of tests.

Several types of soil pressure measuring devices were tested under both aforementioned soil conditions. These included the following instruments: a membrane-type cell similar to that developed by Peattie and Sparrow (1954) with soil pressure-bearing membranes 2.3 and 2.9 mm thick; two modifications of a soil pressure cell with a sound wire type transducer; and a stiff, flat cell with a hydraulic multiplicator and furnished with a bonded soil resistance strain gage transducer. All these instruments were tested under both described conditions; i.e., σ_r/σ_z = 0.57 to 0.53 and σ_r/σ_z = 0.44 to 0.38, and, as mentioned earlier, instrument readings in the absence of soil lateral deformation condition were taken as 100%.

In general, these tests revealed that variations in lateral pressures have a pronounced effect on the performance of some types of soil pressure cells. Characteristic deviations in cell readings when the surrounding soil can expand laterally (e.g., σ_r/σ_z = 0.44 to 0.38) are listed in Table 2-1. The maximum deviation in soil pressure reading, which amounts to approximately 45%, is possibly related to the large area of the side surface of the cells (height-to-diameter ratio approximately 1:5). The insufficient rigidity of the cell's body-membrane system is another possible explanation of the observed phenomenon. It was found that even a small variation in lateral soil pressure exerted upon the side surface of a soil pressure measuring cell, and particularly one having a highly sensitive vibrating wire measuring system, influences substantially the readings taken by this instrument.

It should be noted that in this investigation an attempt to determine the relationship between the rigidity of the membrane-type soil pressure cells and the variation in cell readings with a decrease in σ_r was not successful.

As shown from the data presented in Table 2-1, the soil pressure cell with a hydraulic multiplicator is least sensitive to a variation in lateral soil pressure exerted upon it within limits of the tests in question and lateral deformation of soil in the calibration chamber. This suggests that thin and rigid soil pressure cells with hydraulic multiplicators can be effective in measurements of soil pressures in zones of foundation soil where plastic deformations can be developed.

Large-scale model tests were then carried out to confirm the results obtained from the tests in the calibration chamber. For these experiments a stiff reinforced concrete block 1.10 m in diameter and 0.50 m deep was used (Figure 2-8). The granulometric composition of the founding soil used in these tests was similar to that used in tests in the calibration chamber. Fifteen contact soil pressure cells were installed in the test block and carefully flushed with its bottom. All cells had a hydraulic multiplicator and were furnished with a vibrating wire type transducer. The sensitivity of these instruments is sufficiently high: it is not less than 70 Hz per 0.1 MPa. Furthermore, five rigid flat soil pressure cells were installed in the mass of soil 15 cm below the underside of the test block (Figure 2-9). These cells were placed along the edge of the block, i.e., in the region where the soil is deformed the most under the effects of vertical load and where the lateral pressure associated with soil deformation decreases. All soil pressure cells were then calibrated in the aforementioned calibration chamber.

In addition to contact stresses, settlements of the block was also measured by means of mechanical dial gages. The experiments were conducted in a large

Table 2-1. Effect of Lateral Soil Pressure on Change in Readings of Pressure Cells of Different Designs

Instrument	Rated Modulus of Elasticity (MPa)	Overall Dimensions (mm)		Deviation in Readings from $\sigma_r/\sigma_z = 0.53$ to 0.57 (%)	
				Vertical Stress (σ_z) in MPa	
		Diameter	Height	0.20	0.40
Peattie and Sparrow (1954) type of soil pressure cell with membrane 2.3 mm thick	156	100	12	12	12
Same with membrane 2.9 mm thick	312	100	12	25	14
Soil pressure cell with vibrating wire type transducer developed by Institute Gidroproect, Moscow	196	130	25	24	21
Same, however approximately four times stiffer	800	130	25	16	45
Stiff, flat cell with bonded wire resistance strain gage transducer and with hydraulic multiplicator developed by TSNIISK, Moscow	more than 10,000	120	13	–6[a]	–4[a]

[a]Negative values of readings indicate decrease in cell readings.

Figure 2-8. Test block 1.10 m in diameter with soil pressure cells.

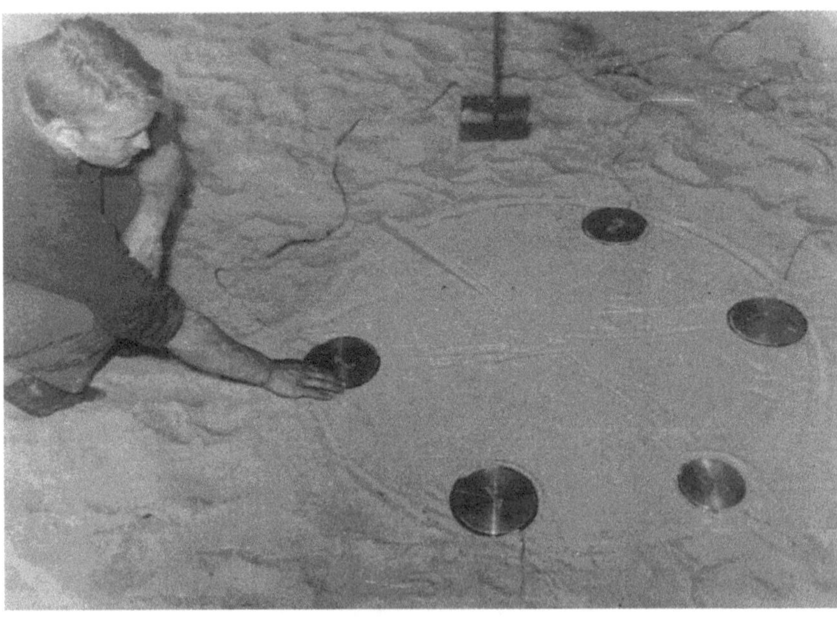

Figure 2-9. Installation of flat soil pressure cells in mass of foundation soil.

concrete bin 5 m wide, 20 m long, and 5 m deep. The load was applied in increments of 0.05 and 0.1 MPa at the bottom of the block. The load was brought up to the level of 0.3 MPa in the first set of loadings, and to 0.5 MPa in the second set.

The results of these experiments are presented in Figure 2-10. Each test point in this figure represents the mean value of 6 (or 3 in the center of the block) readings of soil pressures from cells located at equal distances from the center of the block. As is seen from Figure 2-10, during test loads of up to 0.2 MPa in the first set of loadings, and up to 0.3 MPa in the second set, the readings of pressure cells installed in the mass of soil are very close to those of contact pressure cells.

However, beginning from a pressure of 0.3 MPa and up during the first set of loadings, and 0.4 MPa and up during the second set, the value of soil pressures obtained by the pressure cells installed in the mass of soil begin to lag behind the values measured by the contact pressure cells. At 0.5 MPa pressure, this difference manifests itself especially strongly. Note that the soil pressure measured in the underlying soil and equal to 0.26 MPa at the external pressure of 0.5 MPa is less than the one measured when the external pressure of 0.3 MPa is applied; the latter is equal to 0.27 MPa.

The ratio p_v/p_c in percents, shown in Figure 2-10d, indicates that the difference between p_v and p_c, at least up to an external pressure of 0.3 MPa, is consistent and constitutes approximately 16 to 17%. Here p_v = readings equal to cells installed in the soil mass and p_c = readings by cells installed in the test block. Note that readings of the latter cells are taken as 100%.

The character of test block settlements (Figure 2-10d) and the shapes of soil reaction diagrams 4 and 5 in Figure 2-10c indicate the formation of plastic (shear) zones at the peripheral edge of the test block. This corresponds to loads equivalent to or greater than 0.3 MPa. These experiments indicate that, despite a decrease in soil lateral pressures in the plastic (shear) zone under the test block, no increase in sensitivity of the flat soil pressure cells located in this zone is observed. Quite the opposite, these cells produce progressively smaller reading values beginning with the applied pressure equal to 0.3 MPa and up, which is clear evidence of the decrease in stresses with a mass of soil in this area.

The following basic conclusions have been drawn from the tests described in Chapter 2:

- Rigid, thin soil pressure cells, placed in a mass of soil in close proximity to the underside of the test block, can measure the soil pressure at the block/soil interface with sufficient accuracy at certain loading ranges, both qualitatively and quantitatively. This, however, can change quite dramatically under the effects of relatively higher loadings, particularly in zones of plastic deformations within the founding soil. This zone can develop at the foundation's edges.

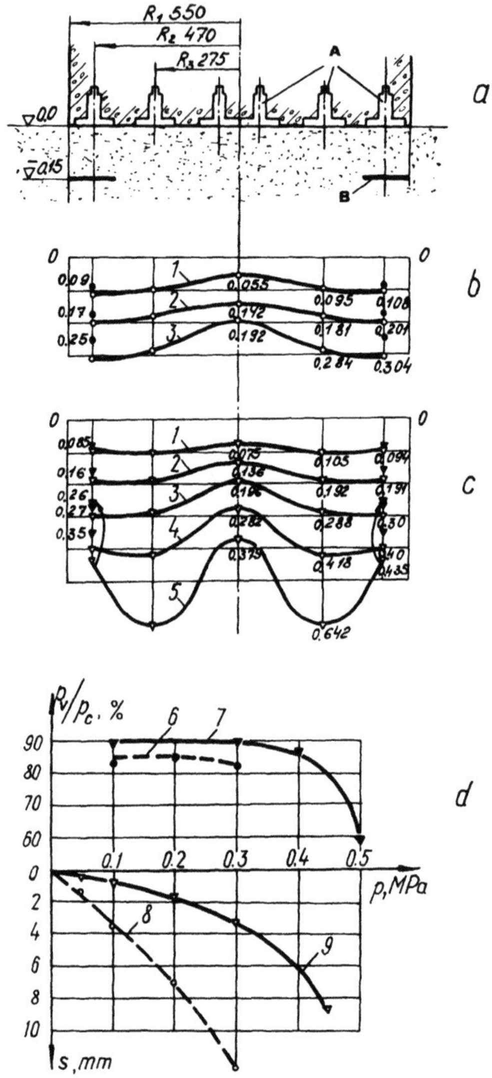

Figure 2-10. Rigid test block. (a) Cross section; (b) soil reaction diagrams in the first set of loadings; (c) soil reaction diagrams in the second set of loadings; (d) diagrams of relative stresses and settlements of test block. A. Contact soil pressures cells with hydraulic multiplicator and sound wire type transducers; B. stiff flat soil pressure cell with hydraulic multiplicator and bonded foil resistance strain gager type transducer. 1 to 5. Diagrams plotted from readings of soil pressure cells correspond to uniform pressures from $p = 0.1$ to 0.5 MPa; 6 and 7. relative pressures p_v/p_c) diagrams of first and second loadings, respectively; 8 and 9. block settlement diagrams during, respectively, first and second stages of loadings. *Note*: Blank test points are related to readings from cells A, and black ones to readings from cells B.

- Measurement of contact soil pressures is better achieved by dedicated contact pressure cells installed in the body of the footing and accurately flush-mounted with the footing's bottom.
- The result of investigations carried out in a calibration chamber indicate that the cells tested in soils able to expand laterally generate an output signal substantially stronger than when the same instruments are tested in soils having no ability to expand laterally.
- The readings of soil pressure cells that are placed in a mass of sand underlying a structure depend on several factors, such as the distance at which these instruments are located from the underside of the structure (in other type of soils, this may work differently); the possibility of formation of plastic (shear) zones, particularly at the edges of the footing; and variation in pressure cell sensitivity with variation in changes of the ratio of soil pressure components (σ_r/σ_z) that is developing in the process of growth in stresses and deformations in the soil mass.
- The stiff, flat, and rather thin soil pressure cells with hydraulic multiplicators that were used in the above-discussed experiments have demonstrated stable performance (in a range of 2% or less) under the relative lateral deformation of soil.

3

Calibration Apparatuses and Calibration Methods of Soil Pressure Cells

3.1 State-of-the-Art Review

An instrument reading is useful only if its correct response to the effects of the pressure it is exposed to is known. This is why prior to installation any soil pressure cell must be properly calibrated; i.e., the relationship between the instrument's output signal and the known applied load must be properly established.

To obtain credible performance in the field, the soil pressure cell should be calibrated under conditions that closely resemble the real conditions in which it is intended to be used. This is why the calibration apparatus (in most cases a calibration chamber) and methods of cell calibration are of paramount importance in the whole process of performance monitoring of large-scale models or real structures. For credible results both the calibration apparatus and calibration method must be developed in a way to avoid or minimize errors, e.g., nonhomogeneity of soil, that can produce adverse effects on cell calibration.

In general, calibration of a soil pressure cell is an expensive and difficult undertaking. The most complex and difficult aspect is the calibration of soil pressure cells placed in a soil mass. The difficulties that can exist in this case are discussed in Chapters 1 and 2. As was pointed out, calibration of contact soil pressure cells that are intended to be used flush mounted with the structure's face exposed to soil pressure differ from calibration of those cells that are to be installed in a soil mass. However, in both cases the uniformity of soil placement within the calibration chamber is a must.

Typically, soil pressure cells are calibrated in the following sequence:

- Initially instruments are calibrated in the calibration chamber by application of hydrostatic or air pressure. At this stage the instrument's physical condition, character of response to the applied pressure, sensitivity, and return to zero reading after the load is removed are checked.

- The pressure cell is loaded in the same calibration device, but this time through a layer of soil. Most commonly this loading is carried out under uniaxial soil deformation conditions; i.e., conditions under which the soil does not deform laterally. From the results of these tests, using fine or medium sands, one can judge with a degree of confidence the instrument stiffness, and the effects of its dimensions and orientation on readings.

As pointed out in Chapter 1, it is important to understand that the relative error (Δ_p) of performance of a thin cell is related to the ratio of deformation moduli of the cell and the soil. Also, in the case of unidimensional compression, the modulus of soil deformation (E_s) increases as the vertical pressure increases. This will be discussed further in this work.

A rigid cell placed in a soil mass will normally respond to soil pressures similar to those found in a stiff foundation/soil interaction. For example, when a foundation model is loaded in a test tank under uniaxial conditions, i.e., where no soil lateral expansion is possible, the soil exhibits a constant pattern of σ_r/σ_z ratio, where σ_r is a function of σ_1 and σ_2 (here σ_1 and σ_2 are principal stresses in the soil mass, as discussed in Chapter 2). In this case the soil modulus of deformation (E_s) increases with each load increment. On the other hand, when similar types of foundations are placed upon soils that can expand laterally, the ratio σ_r/σ_z varies continuously (at least in the case of noncompacted sand), and the soil modulus of deformation (E_s) decreases with increases in load. In other words, the stress distribution under the foundation block is more complex when soils can expand than under uniaxial conditions. This is common knowledge; plenty of related examples are found in standard texts on soil mechanics and foundation engineering.

The aforementioned effects of soil behavior under loads produced by a stiff footing are usually considered in the design of cell calibration devices. The designs of most calibration devices (chambers) used to date have been based on the principle of uniaxial compression. The following is a brief discussion on some of these devices.

A calibration device designed in the form of a cylinder in which soil is subject to uniaxial compression was used by A. Benkelman and R. Lancaster, as discussed by Peattie and Sparrow (1954). The load in this device was applied to the surface of the soil via a rigid concrete slab. As was pointed out in Chapter 2, the soil pressure distribution under a rigid footing is not uniform (Figures 2-5 and 2-6). For example, in compacted soil, the soil reaction has its *minimum* at the center of the footing, but when the soil is not compacted, the soil reaction can reach its *maximum* at the central part of the footing. In both cases the true value of the soil reaction is quite uncertain because of the nonuniformity of soil parameters, in particular the soil density. Because of this, the instrument cannot be properly calibrated in the device. Thus, calibration of pressure cells in the Benkelman-Lancaster device is not reliable and should be avoided.

Peattie and Sparrow (1954), Dergachev (1957), and Trollope and Lee (1961) proposed cylindrical calibration devices with rigid bottoms (Figures 3-1 through 3-3). The calibration load in these devices is applied to the soil surface by liquid or air via a thin rubber diaphragm. As shown in these figures, the structure of these devices and concept of operation are self-explanatory. The advantage of this design is that the depth of soil in the calibration chamber can be adjusted as required. Trollope and Lee (1961) investigated the effects of soil friction about the chamber's wall and concluded that it may significantly affect the stress distribution within the soil mass. The effects of wall friction depend on the ratio of the depth of soil (H_s) to the chamber diameter (D_{ca}). In experiments conducted by these workers, the effect of friction did not manifest itself in the central part of the soil volume when the layer of soil in the chamber was small (say ($H_s/D_{ca} = 1/6$); at $H_s/D_{ca} = 1/3$ the actual stress in the central part of the soil volume constitutes less than 90% of the applied uniform load (p), and at $H_s/D_{ca} = 1/2$ the actual stress in the central part of the soil volume only slightly exceeded 80% of p. Apparently, neither Peattie and Sparrow (1954) nor Dergachev (1957) carried out similar investigations.

Hadala (1967) discusses a calibration chamber developed by the U.S. Army Corps of Engineers that is very similar to that developed by Peattie and Sparrow (1954). This chamber was used to calibrate small (50-mm diameter) cells with

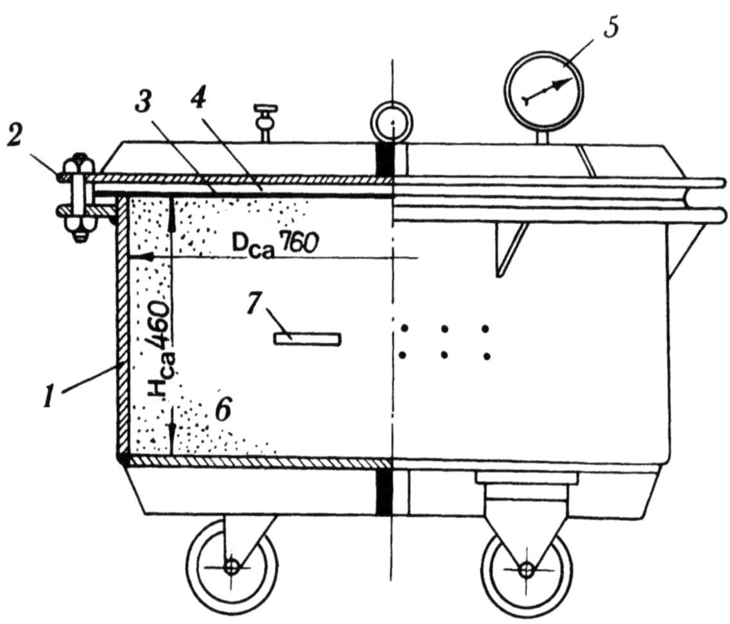

Figure 3-1. Calibration chamber developed by Peattie and Sparrow (1954). Typical section. 1. Cylindrical body; 2. cover; 3. rubber diaphragm; 4. liquid; 5. pressure gage; 6. soil; 7. pressure cell.

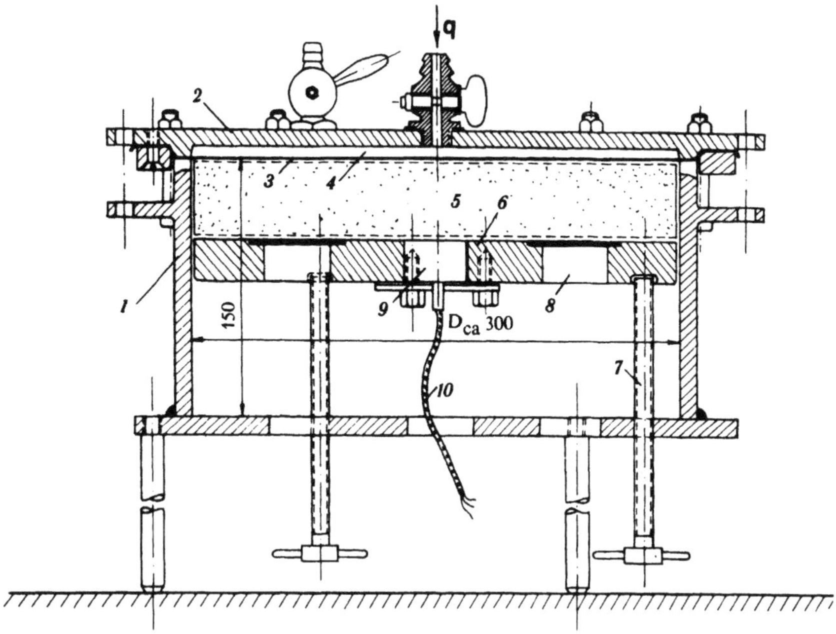

Figure 3-2. Calibration chamber developed by Dergachev (1957). Typical section. 1. Cylindrical body; 2. cover; 3. rubber diaphragm; 4. cavity; 5. soil; 6. steel plate; 7. screw jack; 8. nest for pressure cell; 9. pressure cell; 10. wire.

aspect ratios of 1:9 in both sand and clay. Hadala reports both substantial over- and underregistration ratios and scatter of the data about the mean. According to Hadala, this is basically attributed to the soil placement methods. He concluded that the use of a single soil pressure cell to measure the magnitude of stress in a soil mass with a reasonable degree of confidence is unachievable.

Dunnicliff (1988) suggests that the small size of Hadala's cells aggravated the soil placement effects. He believes that the effects of soil placement would be less severe for more conventional field-sized instruments. However, from our earlier discussion, the effects of wall friction may also be an important factor.

In a calibration device developed by Baranov (1959) the effects of soil friction against the chamber walls are effectively eliminated by the presence of a pressurized "water jacket" installed inside the chamber (Figure 3-4). In this calibration device the load upon the soil can be transmitted either by a piston (via rubber container), or by pumping air or liquid into the system. This apparatus operates on the same principle as a standard triaxial cell for soil testing. To completely avoid the effects of wall friction in this chamber, the tests must be carried out with extreme caution, e.g., the soil should be loosely placed and not compacted inside the chamber. Cyclic application of the load results in both soil densification and increase in the soil modulus of deformation.

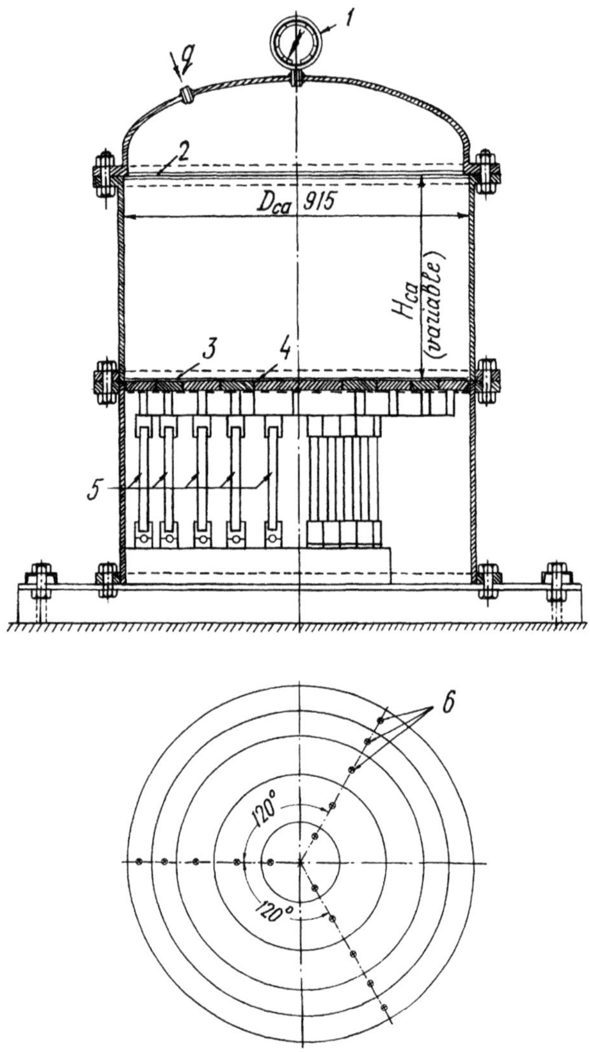

Figure 3-3. Calibration chamber developed by Trollope and Lee (1961). Typical section and plan of composite bottom. 1. Pressure gage; 2. rubber diaphragm; 3. sheet of plastic; 4. base formed from concentric rings; 5. pedestals with strain gages; 6. support points.

We next describe several calibration devices in which, to reduce the effects of wall friction, the walls are designed to contract along with the soil. The apparatus illustrated in Figure 3-5 was developed by Kallstenius and Bergau (1956). It is designed to calibrate the contact soil pressure cells. This device is made from seven individual steel rings connected with each other by a rubberized fabric. The soil in the chamber is loaded by applying a hydraulic pressure. The

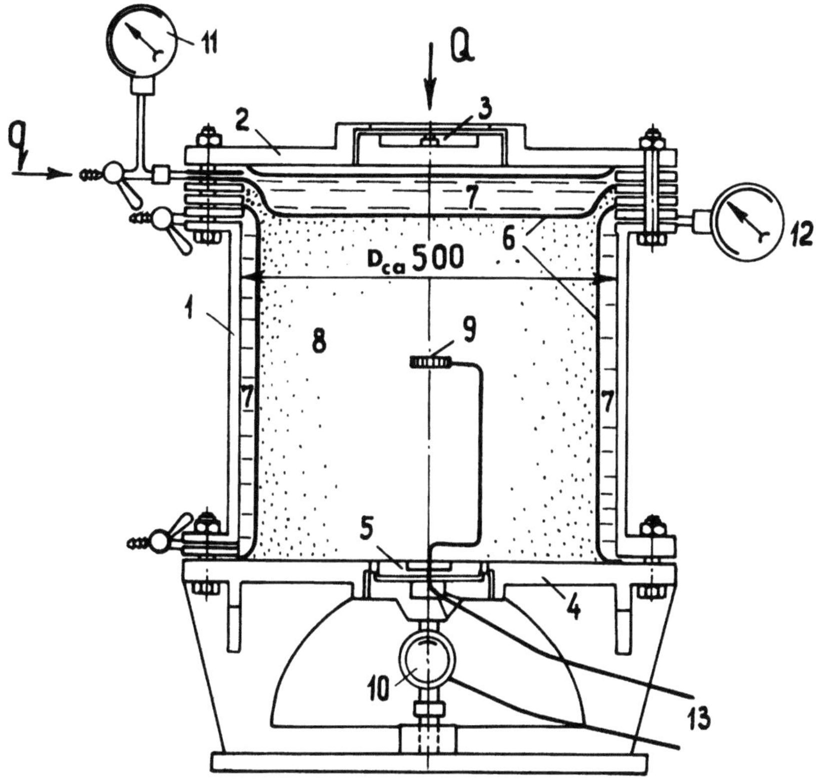

Figure 3-4. Calibration chamber developed by Baranov (1959). 1. Cylindrical body; 2. cover; 3. piston; 4. bottom; 5. conductor; 6. diaphragm and envelope made from thin rubber; 7. water; 8. soil; 9. pressure cell; 10. dynamometer; 11 and 12. pressure gages; 13. wires.

test load (Q) is transmitted through a stiff plate to the soil via a water-filled rubber container. The applied pressure is monitored by a pressure gage. The value and uniformity of axial compression of the chamber's wall are measured by four dial gages installed outside the chamber along its perimeter. The stress and strain of the steel rings forming the chamber are measured by wire resistance strain gages.

In this chamber Kallstenius and Bergau carried out experiments to check the stress distribution along the chamber bottom. Despite a very careful and uniform filling of the chamber with soil, a substantial scatter in data about the mean was observed. Special sensors, 60 mm in diameter, placed at the chamber bottom at a distance of 80 mm from its center, produced data that deviated from the mean in a range from 31% to −20%. The same sensors arranged at a distance of 160 and 220 mm from the chamber's center produced data that deviated from the

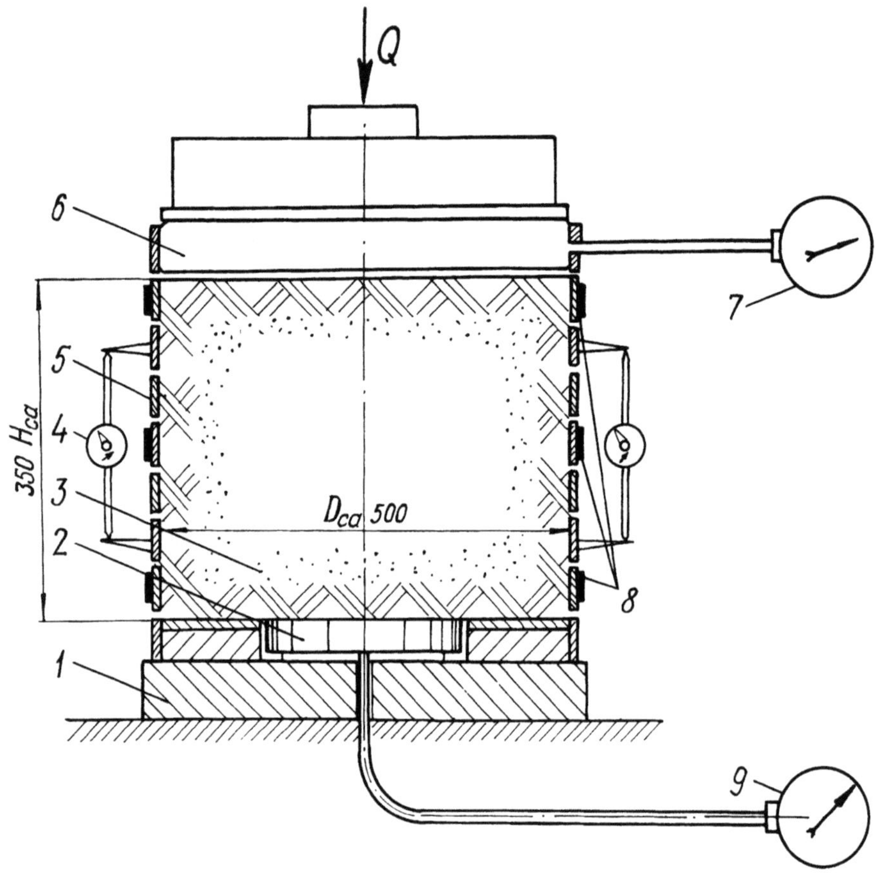

Figure 3-5. Calibration chamber developed by Kallstenius and Bergau (1956). 1. Stiff base; 2. soil pressure cell; 3. soil; 4. strain gage; 5. one of 7 rings forming the chamber; 6. rubber container for water; 7. manometer; 8. wire resistance strain gages; 9. pressure cell manometer.

mean in a range from 36% to −73% and 57% to −22%, respectively. Kallstenius and Bergau linked the scatter in the sensor readings to the effects attributed to wall compressibility.

A calibration apparatus similar to that of Kallstenius and Bergau was designed by Franz (1958). In this device the chamber is comprised of three telescopic sections (rings), where the middle ring is narrower than the lower and upper rings. The test load is applied to the soil via a stiff cover plate.

Kheifitz and co-workers (Kheifitz and Radkevich 1969; Kheifitz 1973) designed a calibration chamber that they called the "large-size pistonless odometer" (Figure 3-6). It differs from the previously discussed calibration devices in the design

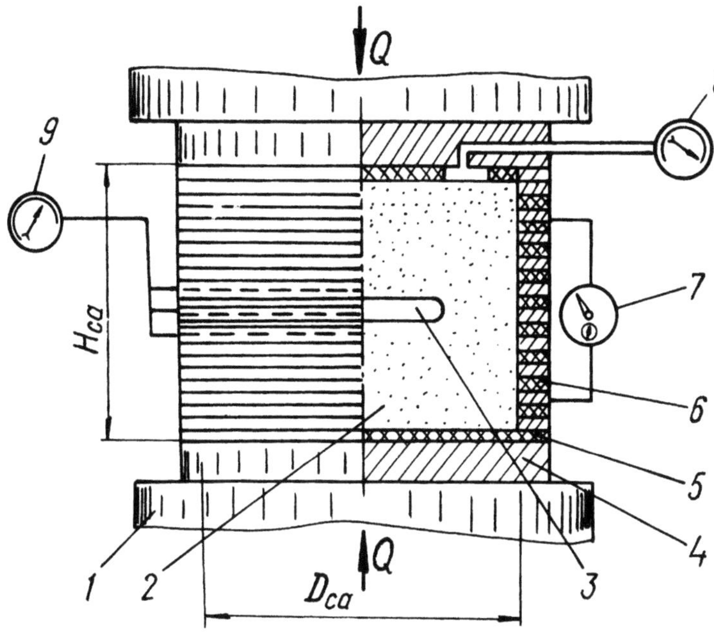

Figure 3-6. Pistonless odometer (Kheifitz and Radkevich 1969). 1. Stiff base; 2. soil; 3. soil pressure cell; 4. stiff bottom; 5. perforated rubber pad with liquid-filled voids; 6. wall comprised of alternating metal and rubber rings; 7. strain gage; 8 and 9. instruments measuring axial and radial stresses in soil.

of the wall, which consists of alternating steel (stiff) and rubber (compressible) rings. The device is rather small; its chamber has a height equal to 360 mm and an internal diameter equal to 350 mm. The test chamber is actually sandwiched between top and bottom stiff plates, and the test load is transmitted to the soil via these plates underlain with thin perforated rubber pads containing liquid-filled voids. According to Kheifitz (1973), a high degree of data uniformity was obtained with this calibration device.

Similar devices 100, 500, and 1300 mm in diameter have been used in Italy for determining the compression parameters of large-grained materials used for the construction of rock-filled dams (Fumagalli 1969).

3.2 Improved Calibration Devices and Investigation of Their Performance

As discussed earlier, the accuracy of soil pressure measurements in both a calibration device and in the field depends on the stiffness of the soil pressure cell, the yield parameters of the pressure membrane exposed to the soil (contact surface), the stiffness of the structural material in which the instrument is installed, and, last but not least, the soil parameters. A characteristic example is illustrated in Figure 3-7.

The soil friction effect on cell performance is not considered in this figure. Hence, the average soil stress (p) at the interface with the upper face of the stiff base (insert) will be equal to the active pressure (q); i.e., $p = q$. However, the contact pressure cell installed in the recess located in the middle of the insert can change this balance in two ways. First of all, if the cell's deformability is higher than that of the material of the base, then under pressure q, the cell's contact surface will deflect below the insert's face. This will result in a pressure distribution as indicated by the dotted line in Figure 3-7. Here the effective pressure exerted upon the cell's contact membrane (p_{min}) is less than the average stress (p). Second, the opposite result can be expected in the case when the cell's deformability is smaller than that of the material of the base (insert). In this case the effective pressure exerted upon the pressure cell (p_{max}) is expected to be greater than the average stress (p). This is represented by the solid line in Figure 3-7. The obvious conclusion from this example is that for better calibration results the deformabilities of both the cell and the insert material must be as close as practical.

Based on these basic principles, Lazebnik and Smirnov (1968) introduced several changes to the conventional calibration chambers that resulted in significantly enhanced performance of such devices. Examples of the calibration devices developed by Lazebnik and Smirnov are depicted in Figures 3-8 through 3-10.

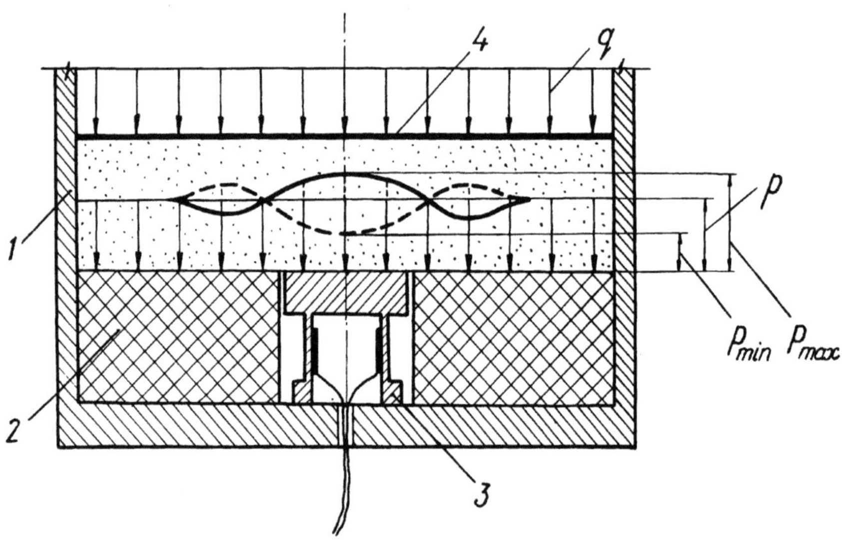

Figure 3-7. Example of effects of stiffness on compatibility between bottom of calibration chamber and contact pressure cell. 1. Chamber; 2. insert; 3. soil contact pressure cell; 4. flexible diaphragm.

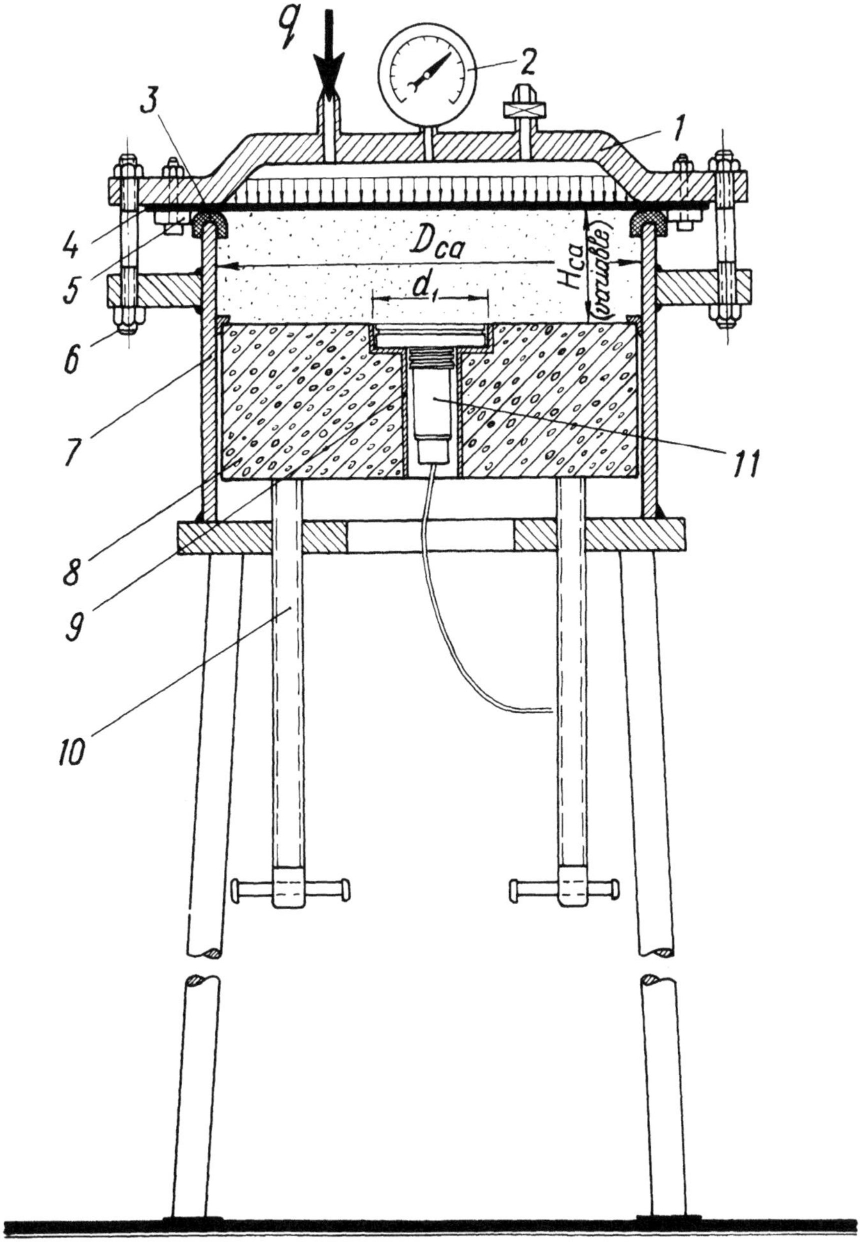

Figure 3-8. Calibration device SCC-475-0.5. Typical cross section. 1. Cover; 2. pressure gage; 3. packing; 4. thin rubber diaphragm; 5. steel ring; 6. bolt; 7. cylindrical chamber; 8. reinforced concrete insert; 9. nest for contact pressure cell; 10. screw jack; 11. contact soil pressure cell.

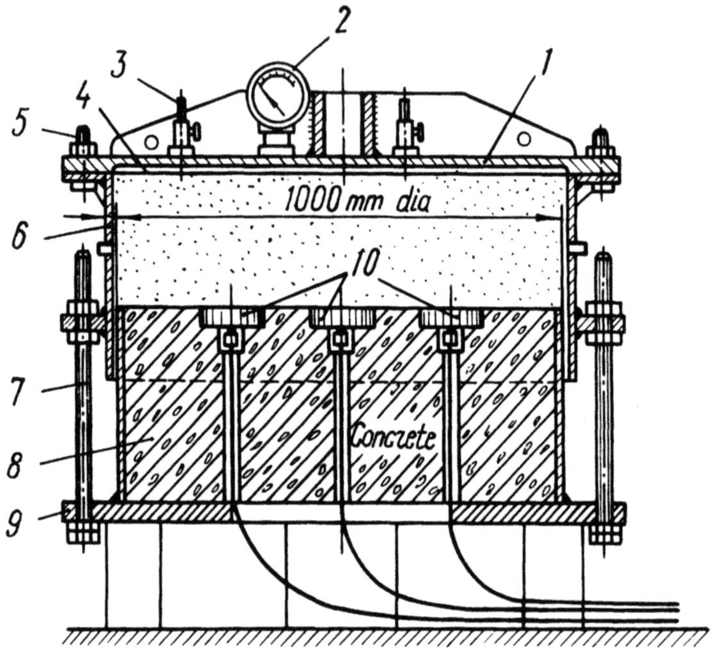

Figure 3-9. Large calibration device LCC-1000-1. Typical cross section. 1. Cover; 2. pressure gage; 3. compressed air nipple; 4. rubber diaphragm; 5. bolt; 6. cylindrical chamber; 7. chamber depth adjusting bolt; 8. concrete in base; 9. steel base plate.

The smaller calibration device (SCC-475-0.5), depicted in Figures 3-8 and 3-10, and the larger device (LCC-1000-1) in Figure 3-9 were developed in the late 1960s. Typical designations follow: SCC-475-0.5 stands for "special calibration chamber," 475 mm in diameter, designed for a maximum calibration pressure of 0.5 MPa. LCC-1000-1 stands for "large calibration chamber," 1000 mm in diameter, designed for a maximum calibration pressure of 1.0 MPa. In both devices the chamber is slightly tapered, e.g., the upper chamber diameter is 2 mm smaller than the bottom diameter, and the internal surface of the chamber is chrome plated and polished to a mirror finish.

The air pressure is applied to the soil via a rubber diaphragm within both devices. The applied pressure is controlled by two standard pressure gages. The chamber's aspect ratio (H_{ca}/D_{ca}) is controlled by a very stiff insert that can be lifted or lowered by four screw jacks, as required. The maximum depth of soil within this chamber is equal to H_{ca} = 170 mm. Because of their simple design, these devices also proved to be very simple to operate—installation of pressure cells, placing and compacting soil, installation and sealing of the chamber's cover, and conducting calibration or investigations do not present any difficulties.

Figure 3-10. Calibration device SCC-475-0.5 with concrete insert raised to its highest position. Shown are four contact pressure cells installed into concrete insert.

Both calibration devices LCC-1000-1 and SCC-475-0.5 are designed and operate according to the same principles.

As has been pointed out, potential errors in calibrating pressure cells in a conventional calibrating chamber are usually caused by soil friction along the chamber's internal surface. Frictional forces "steal" part of the vertical load, which results in a decreased intensity in effective pressure on the bottom (insert) of the device and subsequently on the soil pressure cell(s) installed on it. Soil friction depends on a variety of factors, e.g., the shape and finish of the internal wall surface, deformability of the soil, and others. It is very difficult to estimate wall friction with confidence; therefore, the determination of a soil friction value in a calibration device is usually done experimentally.

Lazebnik (1966) carried out tests to determine the effects of wall friction in the above-discussed calibration chambers (SCC-475-0.5 and LCC-1000-1). Clean, dry medium sand was used in these experiments. The following paragraphs describe the tests that were carried out in calibration chamber SCC-475-0.5.

The effects of soil friction were checked for different pressures that were applied in steps of 0.05, 0.10, and 0.20 MPa upon soils of different densities, e.g., dense and loose. The depth of soil in the calibration chamber was changed three times, obtaining aspect ratios H_s/D_{ca} equal to 0.105, 0.210, and 0.315. The soil pressure was measured by means of four contact soil pressure cells having hydraulic multiplicators and a strain gage measuring system. The stiffness of these cells was compatible with that of the material of the insert in which they were installed. This eliminated the possibility of distortion in the soil pressure measurements. Initially, the pressure cells were calibrated by compressed air, and the subsequent negative balance between the air and the measured soil pressures was assumed to be attributable to soil friction on the chamber wall.

The results of these experiments, in the form of $p_m/p = f(H_s/D_{ca})$, are illustrated in Figure 3-11. Here p_m and p are, respectively, the measured soil pressure and the applied pressure. All tests were duplicated and the soil pressure in each test was measured by four pressure cells, as illustrated in Figure 3-10. Hence, each experimental point essentially represents the averaged data obtained from eight measurements.

As is indicated in Figure 3-11, the effect of soil friction in calibration device SCC-475-0.5 for compacted sand is negligible (approximately 1.6% of total pressure) and for the loose sand it constitutes approximately 7.6%. The latter is most likely attributed to higher deformability of the loose sand. It must be noted that repeated loadings of the same sand produced substantially different results in comparison to its first loading. This is discussed in detail later in this book. Later on, similar tests were carried out to the maximum load of 0.35 MPa. These tests produced results that are very similar to those illustrated in Figure 3-11. These tests proved that the effect of soil friction in a calibration chamber can be effectively reduced by introducing slightly tapered and highly smooth walls.

The contact soil pressure cells that were used in the above investigations are of particular interest. They are very stiff cells with hydraulic multiplicators and a resistance strain gage system designed for a maximum load of 0.5 MPa. A typical cross section of this instrument is depicted in Figure 3-12. Each cell was calibrated by loads applied both by the air and mechanically by the hydraulic press. Both produced virtually identical linear calibration characteristics.

Numerous tests have been carried out to determine the potential effect of wall

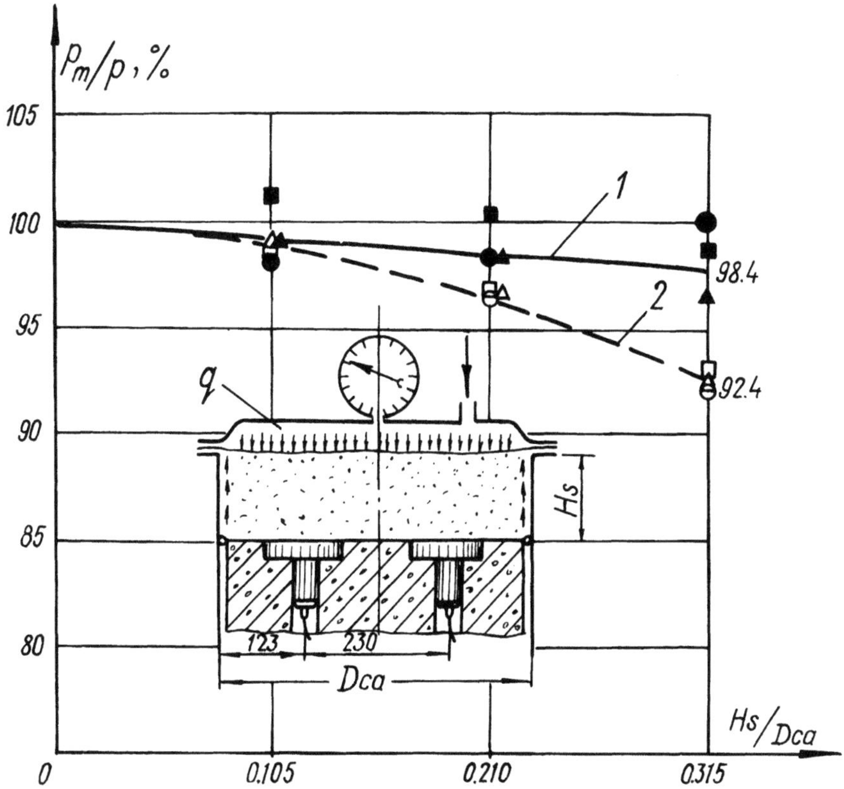

Figure 3-11. Effects of soil friction in calibration device SCC-475-0.5. 1. Loading of compacted sand; 2. loading of loose sand.

friction on soil pressure measurements by the aforementioned cells at the top of the insert and, in particular, at its central area. This pressure cell was loaded in SCC-475-0.5 by air pressure and through a layer of sand. In each series of experiments the load was applied in steps of 0.025 and 0.05 MPa, and the test results (statistical distribution) were processed by using standard equation (3-1).

$$\Delta_p = K\hat{\sigma}_x = K\sqrt{\frac{\Sigma(x_i - \bar{x})^2}{n - 1}} \tag{3-1}$$

where

Δ_p = measurement error

K = a coefficient depending on Δ_p distribution law; K = 2 according to test results

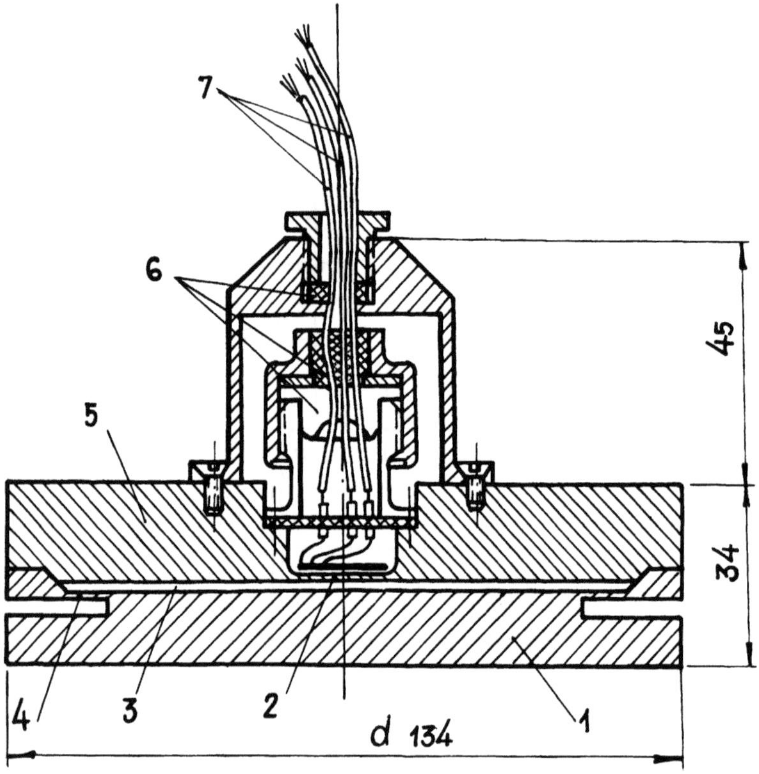

Figure 3-12. Typical cross section of soil pressure cell used in calibration chamber SCC-475-0.5. 1. Contact membrane; 2. measuring membrane; 3. liquid; 4. ring membrane; 5. body; 6. sealing of inner cavity of instrument; 7. transducer wiring.

$\hat{\sigma}_x$ = estimated measurement error

x_i = one measurement

$\bar{x}$ = arithmetic average of the measurements

n = number of measurements.

The statistical analyses of the data obtained from the experiments lead to the following conclusions:

- The effect of soil friction in calibration chamber SCC-475-0.5 is negligible; e.g., at the central part of the insert's face the deviation of the measured soil pressure from that applied by compressed air varied from 1.14 to 0.33%.
- No simple relationship exists between the cell readings and the condition of load application; e.g., in the case of load transfer through the layer

of soil, the cell's signal can be either slightly lower or slightly higher compared to that when the load is applied directly to the cells.

- The insignificant effect of soil friction in calibration chamber SCC-475-0.5 makes it a reliable means for calibrating soil pressure cells.

A follow-up investigation has indicated that the soil pressure within the soil mass placed into SCC-475-0.5 is distributed quite uniformly. The stresses within the soil mass have been measured in SCC-475-0.5 at three levels; as indicated in Figure 3-13. At each level stresses were measured at 20 locations. A small, symmetrically placed soil pressure cell was used to carry out the measurements. As seen in Figure 3-13, soil friction manifests itself only in close proximity to a chamber's wall. And again, it has no influence on the readouts of a cell located closer to the central part of the calibration device. From the soil pressure diagrams depicted in Figure 3-13, it follows that a soil pressure cell with maximum diameter equal to 136 mm placed in the middle part of calibration device SCC-475-0.5 is not affected by soil friction while exposed to calibrating pressures.

Calibration of a pressure cell intended for use within a mass of soil is more difficult than calibration of a contact pressure cell. This is associated with not only the wall friction but also with cell performance within the mass of soil and soil deformability in both real conditions and in conditions that exist in calibration devices.

As pointed out by Kheifitz (1973), the different soil deformation conditions that can exist between the cell placed within a mass of soil in close proximity to the underside of the stiff footing and the footing itself, and below the cell, where deformation of soil is unrestricted, should be considered during the calibration process. An example is illustrated in Figure 3-14. From this figure it follows that if the pressure cell during the experiment or field monitoring program is located with its pressure-bearing membrane facing the stiff footing (Figure 3-14a), then in the calibration device, to simulate the real conditions, it should be loaded facing the device's stiff bottom (Figure 3-14b). Accordingly, if the direction of the active loading pressure with respect to the cell's pressure-bearing membrane is reversed, then the cell should be placed under the footing as depicted in Figure 3-14c and calibrated in the calibration device as indicated in Figure 3-14d.

Lazebnik (1975) proposed modifications to the calibration device that allow for the development of a uniform symmetrical stress state within the volume of soil in the calibration chamber. The conceptual sketch of this improved calibration apparatus is depicted in Figure 3-15. The principle of operation of this device is self-explanatory from this figure. To reduce the effects of wall friction, the chamber internal surface is mirror-finished as in chamber SCC-475-0.5. The chamber's height-to-diameter ratio does not exceed 1:3. [Dunnicliff (1988) suggests that for a reliable outcome the chamber should be at least three times, and preferably five times, the diameter of the cell.] It should be noted that calibration

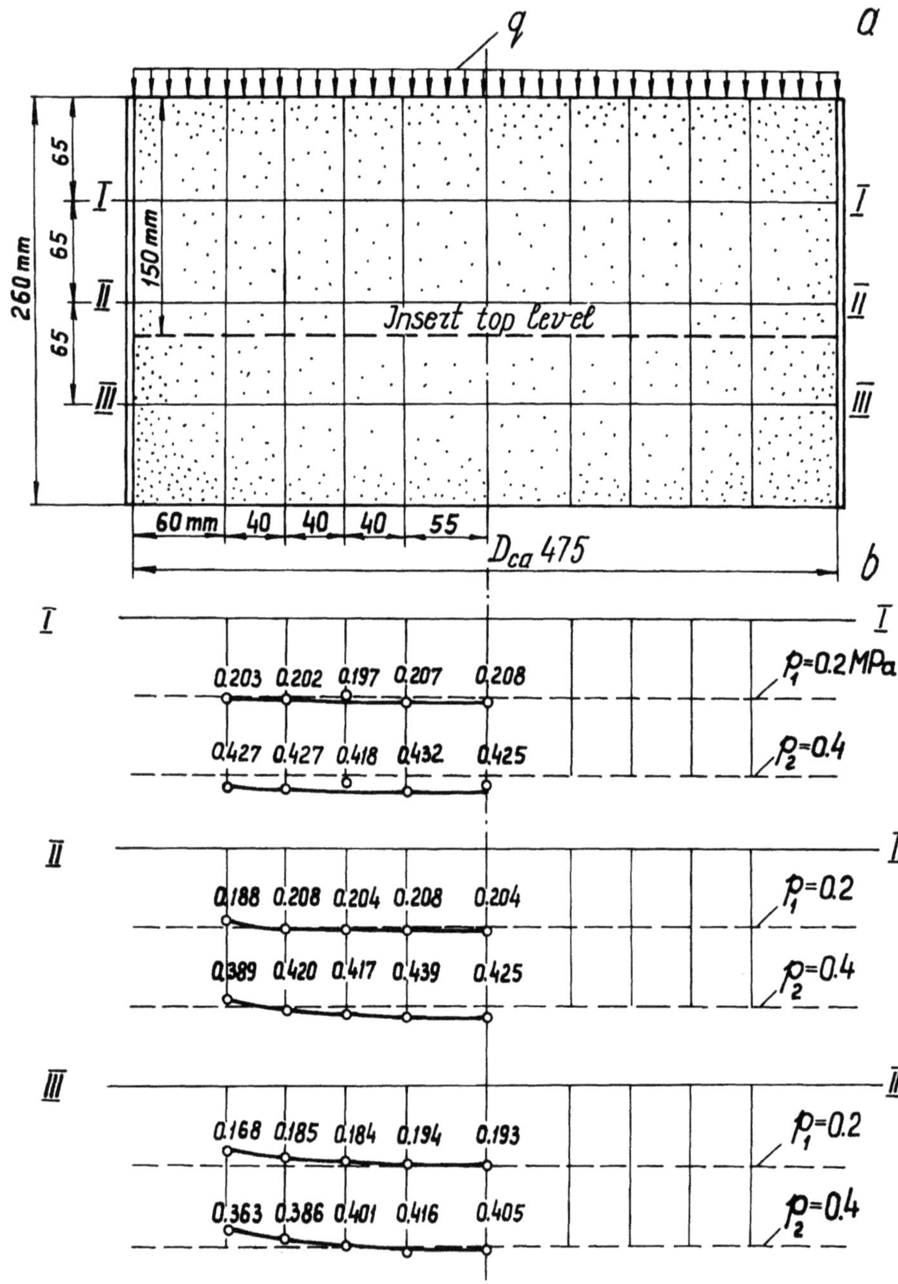

Figure 3-13. Soil pressure distribution within mass of soil. (a) Indication of levels within SCC-475-0.5 at which soil pressure measurements have been taken; (b) soil pressure distribution.

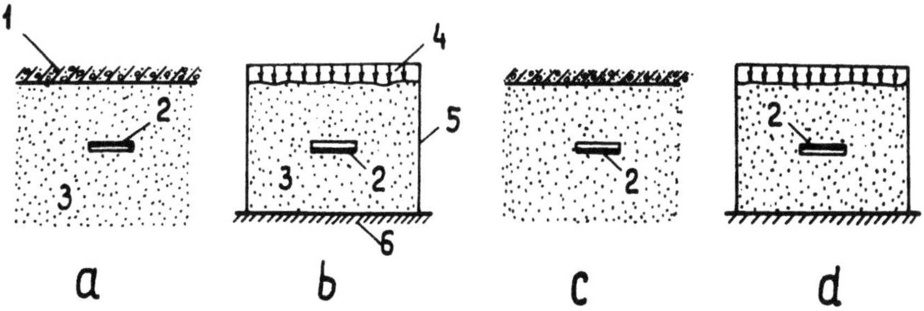

Figure 3-14. Orientation of pressure-sensitive surface of soil pressure cell in mass of soil with respect to direction of active load during experiment or field monitoring program and in calibration chamber. 1. Footing; 2. cell's soil pressure bearing; 3. soil; 4. loading pressure in calibration chamber; 5. calibration chamber; 6. stiff bottom of calibration chamber.

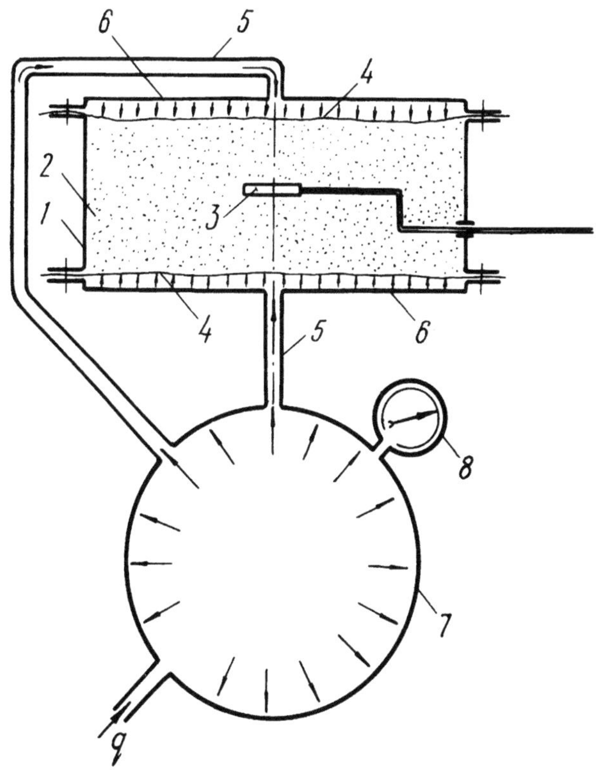

Figure 3-15. Calibration apparatus for application of uniform symmetrical load upon soil pressure cell in mass of soil. Conceptual sketch. 1. Wall; 2. soil; 3. pressure cell during calibration; 4. flexible diaphragm; 5. pressure pipe; 6. cover; 7. receiver; 8. pressure gage.

of the pressure cell in this device, if required, can be carried out by unidirectional loading, as is done in the conventional calibration chamber.

Experiments on two types of stiff cells having soil pressure-bearing membranes only on one side of the cell (nonsymmetrical cell) were carried out in the two-sided loading chamber. It was found that the readings obtained in the double-sided chamber are approximately 3% smaller than those obtained in a conventional calibration device. It was concluded that the difference in cell readings is not so much the effect of the load application, but rather a manifestation of nonsymmetrical cell orientation effects when loaded from both sides.

To conclude this chapter, it must be pointed out that if measurement accuracy must be maximized, then each cell should be calibrated in large enough calibration chambers and both soil placement inside the chamber and cell installation procedures should closely resemble the intended field or experiment methods. Cell calibration is an important, as well as expensive and difficult, undertaking and therefore must be given maximum attention. For more information on calibration chamber design and calibration of soil pressure cells, the reader is referred to works by Hadala (1967), Bozozuk (1970, 1972), Selig (1980), and Alberro and Borbon (1985).

4

Stiff Contact Pressure Cells

4.1 Stress-Strain Measuring Systems: General

Contact soil pressure cells belong to a group of electrical instruments that include electrical transducers of miscellaneous designs. In the past three or so decades numerous types of electrical transducers and associated data acquisition systems have been introduced to the market. Comprehensive reference to these instruments are given in Lion (1959), Perry and Lissner (1962), Prensky (1963), Stein (1964), Cerni and Foster (1965), Norton (1969), Arthur (1970), Considine (1971), Herceg (1972), Spitzer and Howarth (1972), Wolf (1973), Carlson (1975), Dunnicliff (1988), and in numerous works by engineers and scientists in the former USSR. References to these latter works are given elsewhere in this book.

The electrical transducers that are used in geotechnical instruments, and in soil pressure cells in particular, are basically of two designs. The first group includes transducers with electrical resistance strain gages in which the electrical resistance, and therefore the electrical conductivity properties, changes in direct proportion to a change in length. This is usually expressed as $\Delta R/R = (\Delta L/L)K$, *where* ΔR = resistance change, R = original resistance, ΔL = change in length, L = original length, and K = gage factor; in most practical cases $K = 2$ for standard strain gages.

Five basic types of electrical resistance strain gages have been used in resistance strain gage transducers: the bonded and unbonded wire, bonded foil, semiconductor, and weldable electrical strain gages. The first three types of strain gages are used most often in geotechnical instruments. A brief description of the five types follows.

A bonded wire strain gage is normally fabricated with fine copper-nickel or nickel-chromium wire looped back and forth and bonded to a thin elastic pad, e.g., a thin paper or plastic film, which in turn is bonded to the working member located within the transducer body. A bonded foil strain gage is composed of a thin foil of resistance alloy, bonded to a thin plastic film that, similarly to the bonded wire transducer, is bonded to the working member located within the transducer. The foil is usually photofabricated and etched or die-cut during manufacture to produce grid patterns of varying designs.

In the unbonded wire strain gage a fine wire is looped around two sets of electrically insulated posts that are attached to the working member of a transducer. This latter type of strain gage is, in general, less robust and less reliable than the bonded gages and is generally rendered an obsolete technology; with the advent of improved bonding compounds, unbonded wire strain gages have become less common.

Semiconductor resistance strain gages use semiconductor crystals of silicon or germanium to transform changes in the crystal's strain into proportional changes in its electrical resistance. The gage factor (K) of these crystals varies from 50 to 200; this makes this type of gage much more sensitive than other types.

The weldable resistance strain gages are those in which a resistance element, e.g., a bonded wire or foil gage, is permanently attached to a thin stainless steel mounting flange. Semiconductor resistance and weldable resistance types of gages are more suitable for investigation of concrete and steel structures and less suitable for use in geotechnical instruments. For details on these types of gages the interested reader is referred to Wnuk (1981). Also, for a detailed discussion on all five types of strain gage systems, the reader is referred to the works mentioned at the beginning of this section.

The output from an electrical strain gage system is normally measured using a Wheatstone bridge circuit. The system network can include a quarter, half, or full bridge system. The principles of these systems and their advantages and disadvantages are discussed in Dunnicliff (1988).

The second group of transducers used in geotechnical instruments is based on vibrating wire principles. The concept includes the use of a steel wire that is clamped at its ends and tensioned so that it is free to vibrate as a piano string at its natural frequency. The vibration varies with the wire tension, which changes due to relatively small movements between the two clamped ends. One can therefore use the wire as a strain gage by plucking the wire, measuring the natural frequency, and relating frequency change to strain. The wire is plucked magnetically by an electrical coil located near the wire at its midpoint. This same coil or a special instrument is used to measure the frequency of vibration. The relationship between strain (ε) and wire frequency (f) is expressed as (Hawkes and Bailey 1973)

$$\varepsilon = k(f^2 - f_0^2) \tag{4-1}$$

where

$$k = 4L^2\rho/Eg$$

and

f and f_0 = wire natural frequency and the initial frequency, respectively, in $\sec^{-1}$

L = length of vibrating wire in inches

ρ = density of the wire material in lb/in^3

g = gravity acceleration in $in./s^2$

E = modulus of elasticity of the wire in $lb/in.^3$.

The equation for the frequency of a vibrating wire in terms of the wire stress is expressed as

$$f = 0.5L(\sigma g/\rho)^{0.5} \tag{4-2}$$

where σ is the stress in the wire in $lb/in.^2$.

Historically, the major disadvantages of vibrating wire transducers have been wire corrosion, creep under permanent tension, and slippage at clamping points, all of which result in a reduction in the vibrating frequency. Usually it is manifested as a drift of a zero reading. Tharp (1986) reports on measurements of zero readings made 18 months after original calibration on 16 instruments prior to their installation. The change in zero reading varied from 0.06% to 2.58% with the average value approximately equal to 0.54%. To reduce the potential to zero drift Bordes and Debreuille (1985) recommend that the wire tension not exceed 9 to 13% of the wire material yield strength.

Dunnicliff (1988) suggested that the following actions are highly desirable to reduce drift of the instrument's zero point:

- use of appropriate wire attachment procedures at clamping points; this can minimize the potential for corrosion
- adherence to maximum tension limit
- appropriate aging of instrument by thermal and/or strain cycling.

4.2 Developing of Contact Pressure Cells

Cell design, fabrication, calibration, and installation procedures can all contribute to the potential errors in measurements of soil pressures. In this section and elsewhere in this book the term **error** is used to define the deviation between the measured value of soil pressure and the true one, i.e., mathematically it is equal to accuracy. In general, errors can be a result of incorrect cell design, use of inappropriate or faulty materials, corrosion or temperature effects, improper calibration, careless transportation and storage, incorrect installation techniques, misrecording, and other reasons.

In the case of contact pressure cells it is important that the deformability (compressibility) of both the cell and the structure in which the cell is installed are equal or close to equal. As discussed in Chapter 3, this will ensure that the measured pressure (p_m) is equal to the effective one (p).

For example, the deformability of the cell depicted in Figure 3-12 is equal to 0.2 micron at pressure of 0.1 MPa. This is the approximate deformability of a 34-mm-thick layer of concrete, which is equal to the height of the instrument in question. Carlson (1984) suggested that the composite modulus of elasticity of the pressure cell must be at least one-half of the concrete. The thickness of the cell measuring membrane is also depicted in Figure 3-12 and is equal to 0.45 mm. The cell was designed for use in a calibration chamber under carefully controlled laboratory conditions. Its 21-mm-thick soil contact membrane was glued to the instrument body by special high-strength, epoxy-based adhesive.

Subsequently, the cell shown in Figure 3-12 was modified for use in large-scale models and laboratory experiments. This modified cell is shown in Figures 4-1 and 4-2. In this version the soil contact membrane is welded to the cell body. All basic parts of this cell are fabricated from stainless steel. The cell is fitted with a special device that allows its hydraulic cavity to be filled with a special liquid under vacuum. This is done reliably in a special unit without compromising the state of vacuum within the cavity. Typically, the calibration curve of this type of cell is linear, and when calibrated by soil the relative deviation from linearity, Δ_p (error), does not exceed 2%.

Generally, the instrument is fabricated to satisfy the particular test conditions. Its typical specification is given in Table 4-1. Sometimes, e.g., in laboratory scale model tests, it is necessary to measure very low soil pressures, say to 0.01 MPa and less. In this case, fabrication of pressure cells having hydraulic multiplicators is a very difficult undertaking. Besides, the very thin working (internal) membrane of this instrument does not provide for effective attachment and use of the electric resistance gages that are attached to it. This adversely affects the instrument's sensitivity.

In the past, to overcome this problem, investigators used a system comprised of two cylindrical elements with electrical strain gages attached to their sides, as indicated in the conceptual sketch (Figure 4-3). The bottom of the internal cylinder performs as a soil pressure-bearing membrane. To achieve the required sensitivity of the instrument, the internal cylinder is fabricated from an appropriate plastic material that possesses the required elastic properties and does not creep under the soil pressure of design magnitude.

A practical example of this type of contact soil pressure cell is illustrated in Figures 4-4 and 4-5. This instrument was developed by Chernyshova (1963) and

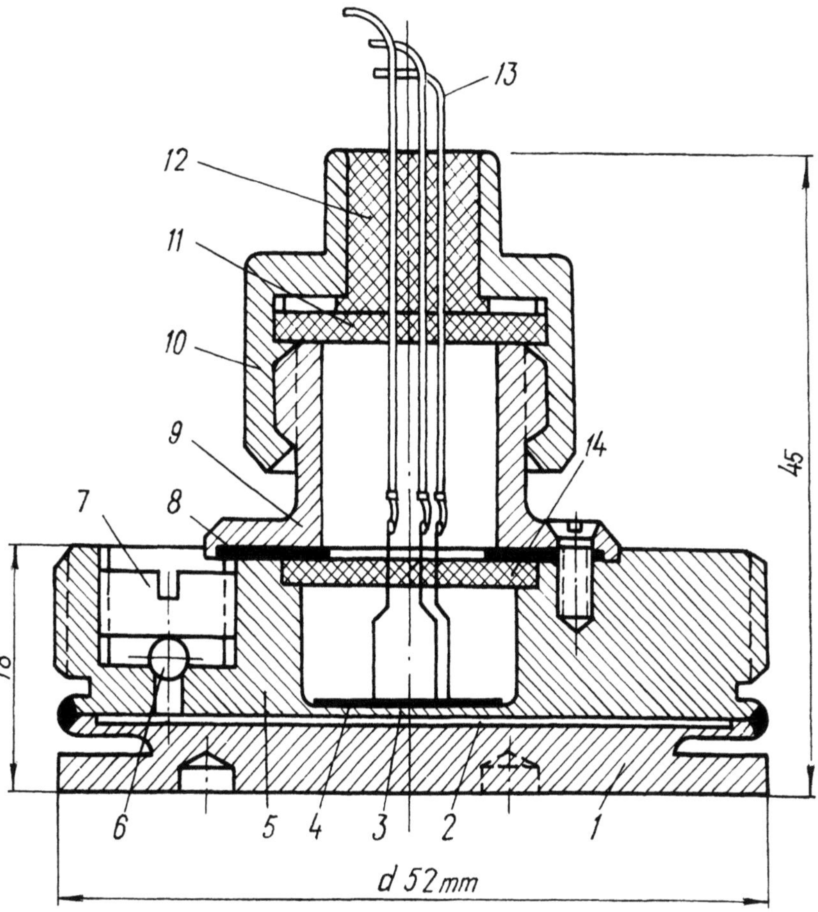

Figure 4-1. Small, stiff contact soil pressure cell. Typical cross section. 1. Contact membrane; 2. hydraulic cavity; 3. measuring membrane; 4. bonded foil strain gage; 5. body; 6. seal; 7. blunt bolt; 8, 11, and 12. sealing; 9. connecting detail; 10. coupling nut; 13. wires; 14. insulator.

this writer in the early 1960s. It was used for the investigation of a lateral soil pressure acting against a large-scale model of a steel sheetpile wall (consult Chapter 11). The inner component of this instrument, exposed to soil pressure, was fabricated from a special plastic material that, at room temperature (20°C) does not creep under pressures in a range of up to seven MPa. The system was joined together and tightened by a centrally located bolt (6 on Figure 4-4).

The plastic material used for fabrication of the internal piston had a modulus of elasticity of 3000 MPa. The diameter of the soil contact surface was 31 mm, and the sensitivity of the strain gage circuit with respect to relative deformation

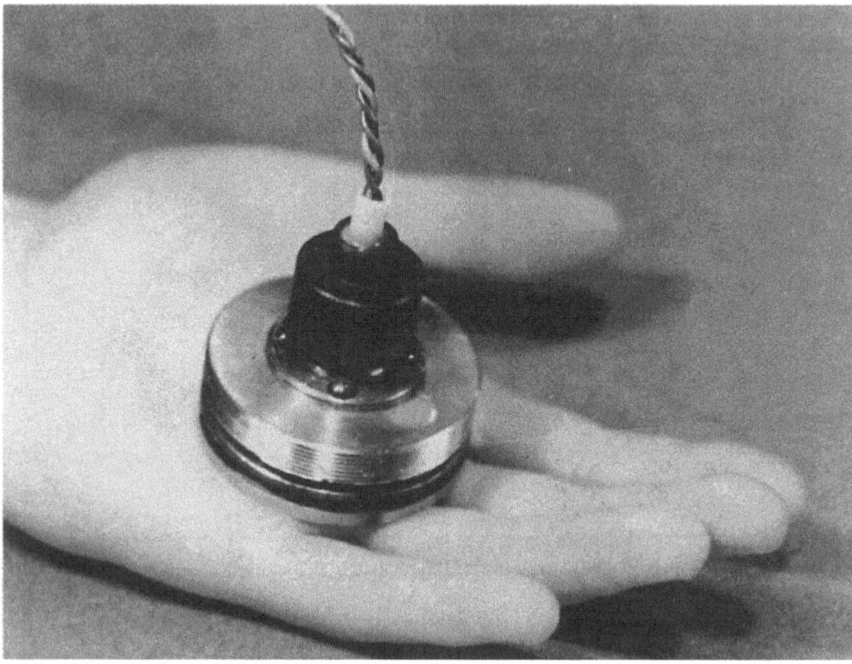

Figure 4-2. General view of contact soil pressure cell depicted in Figure 4-1.

was equal to 0.25×10^{-5}. The imbalance of the Wheatstone bridge was reached at a soil pressure equal to 5.70×10^{-5} MPa. The calibration curves of this instrument are linear when loaded by the compressed air and loose and compacted sands. Details on the Wheatstone bridge are found in Dunnicliff (1988).

In general, bonded electrical resistance strain gages are robust, easy to install, and suitable for monitoring structural dynamic performance. However, despite the advent in the development of bonding compounds used for installation of

Table 4-1. Typical Specification of Contact Soil Pressure Cell Depicted in Figures 4-1 and 4-2 (after Lazebnik 1982)

Maximum (Limiting) Pressure Measured by Pressure Cell (MPa)	Rated Modulus of Elasticity of Pressure Cell Body (MPa × 10^4)	Sensitivity per 10 Relative Deformation Units (r.d.u.), % of Upper Limit of Range
0.1	0.235	1.00
0.3	0.55	0.60
0.5	1.20	0.60
1.0	3.00	0.60
2.0	6.65	0.60
3.0	9.45	0.60
5.0	13.10	0.55

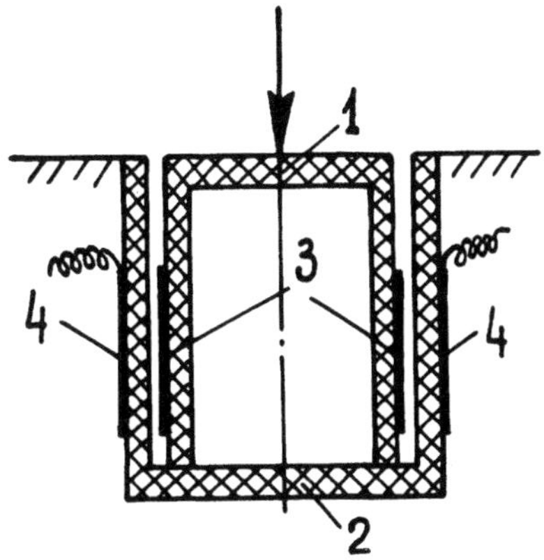

Figure 4-3. Conceptual sketch of double-cylinder contact soil pressure cell. 1. Inner cylinder; 2. outer cylinder; 3 and 4. resistance strain gages.

electrical resistance strain gages, these bonded gages are still subject to some creep when stressed during an extended period of time. For this reason, transducers that use bonded resistance strain gages are usually employed for comparatively short-term investigations, e.g., in a high-humidity environment for no longer than 3 months; the readings taken beyond this period of time are usually regarded as unreliable.

Bonded resistance strain gage transducers also prove to be very sensitive to moisture effects, which usually results in substantial and uncontrollable errors in instrument readings. However, the use of waterproofing materials and coatings can be effective in preventing penetration of moisture inside the instrument. Additional limitations of electrical resistance strain gages are their potential errors attributed to temperature effects, deficiencies of electrical connections, and relatively low electrical output.

To overcome some of these problems the unbonded (elastic) wire strain gage system can be used in contact soil pressure cells. Such instruments were developed by Carlson Instruments, Campbell, California. These instruments have a hydraulic multiplicator containing a cavity filled with mercury. The thickness of the mercury fill in these instruments is less than 0.3 mm. The instrument has a composite elastic modulus of deformation of at least one-half that of concrete. Carlson's pressure cell is described in detail in the company's commercial brochure and in Dunnicliff (1988). Carlson (1978) describes laboratory tests in which these instruments were used. The worst error registered during these tests was 12% of the major principal stress, and this only happened when the load was applied more quickly than would be the case in a real situation.

Obviously, unbonded electrical resistance strain gages do not have the creep problems that are characteristic of the bonded types; otherwise, the advantages and limitations that are typical of bonded strain gages are also characteristic of

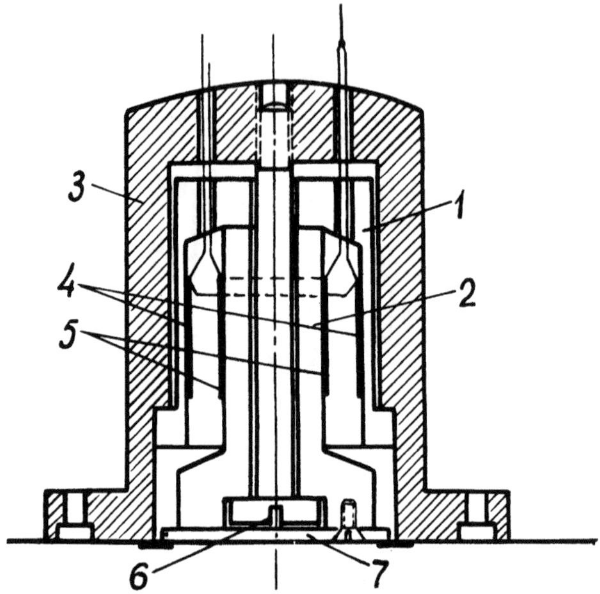

Figure 4-4. Typical cross section of a contact pressure cell for measuring soil pressures of small magnitude. 1 and 2. Respectively, outer and inner components; 3. steel body; 4 and 5. electric resistance strain gages; 6. tension bolt; 7. steel cover serving as contact membrane.

the unbonded electrical resistance strain gages. The advantages and disadvantages of both types of gages are discussed in detail by Renski et al. (1977), Dunnicliff (1988), and others.

Contact (and other) soil pressure cells with vibrating wire type transducer cells have been extensively used since the 1960s. It should be noted that, in general, the vibrating wire (electroacoustical method) has been known and used for an extended period of time. For example, in Russia, Davidenkov (1933) pioneered

Figure 4-5. Different stages of assembly of contact soil pressure cell depicted in Figure 4-4.

the development and use of electroacoustical (vibrating wire) devices for the measurement of soil deformation.

The early experiences in Russia with the use of the instruments based on the electroacoustical method have been summarized by Uginchus and Bombchinsky (1954). These workers indicate that instruments based on the vibrating wire principle can be successfully used for field monitoring performance of miscellaneous hydraulic structures, e.g., dams, spillways, etc. They also conclude that the correctly developed and installed vibrating wire transducer can reliably perform for an extended period of time without significant distortion of readings. The disadvantage of these instruments, however, is that they cannot be used to measure high-frequency dynamic strains.

Lazebnik and his co-workers have developed a family of contact soil pressure cells that for more than 30 years have been successfully used for laboratory and field investigations of soil/structure interaction and for long-term monitoring performance of very important structures, e.g., large industrial and office buildings, nuclear power stations, etc. A representative example of the instrument used for measurements of contact soil pressure at the concrete footing/soil interface is illustrated in Figures 4-6 and 4-7. This particular pressure cell is abbreviated SDKS-1-0.5 (Model 1), which in Russian stands for "welded, vibrating wire pressure cell for measurement of contact soil pressures, with pressure limitation 0.5 MPa." In this instrument, the tension of the wire decreases with an increase in soil pressure. The design of this instrument is self-explanatory from Figure 4-6.

The tension in the vibrating wire depends on deflection of the measuring membrane, which is caused by the pressure from a special liquid filling the hydraulic cavity. This pressure is developed in the process of soil/structure interaction and is delivered to the liquid by a stiff contact membrane. Vibration of the wire is excited by the electromagnet, which also serves as the receiver of the induced electroacoustic signals. Long-term stability of cell performance is achieved by heat treatment of the pressure cell parts and by reducing the depth of the hydraulic cavity to 0.4 mm.

As is seen in Figure 4-6, the contact membrane is welded to the cell body. This has been done in an environment of inert gas (argon) using special alloy rods. The cavity is filled, under deep vacuum conditions, with a liquid having a small coefficient of thermal expansion. The body parts and contact surface of this cell are fabricated from stainless steel; the other parts are made from low-carbon steel and plastic materials (enclosure for electromagnet coil).

Calibration of these cells by air pressure and pressure exerted by sand of different densities has shown that calibration curves are practically linear for this cell. The sensitivity of the SDKS-1-0.5 cells is approximately 0.0015 MPa per 1 Hz. The performance stability of these instruments was extensively tested under laboratory and field conditions. A total of 18 pressure cells were tested in the laboratory during a period of 4 months. Readings were taken at the same tempera-

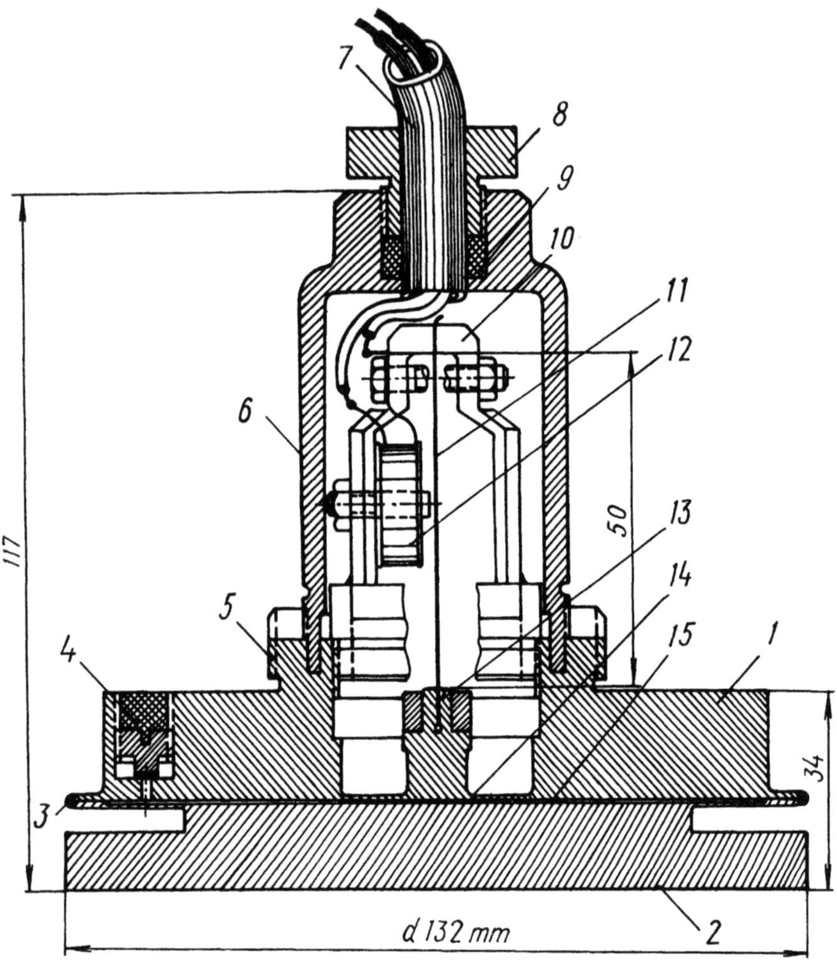

Figure 4-6. Contact soil pressure cell SDKS-1-0.5. Cross section. 1. Body; 2. soil pressure-bearing membrane; 3. circumferential weld; 4. seal; 5. thread; 6. protection cylinder; 7. double-wire cable; 8 and 9. cable sealing; 10. wire upper holder; 11. steel vibrating wire; 12. electromagnet; 13. wire lower holder; 14. measuring membrane; 15. hydraulic cavity.

ture, 23.5°C. The maximum deviation in cell reading constituted just 2 Hz per month, which corresponds to a stress of 0.003 MPa. The mean deviation in readings taken from all 18 pressure cells did not exceed 0.7 Hz (Lazebnik, 1966).

During a year-long operation of these cells in the field, where the structure was founded upon high-water content foundation soils (clayey and sand), the maximum variation in cell reading (with no increase in load on the structure)

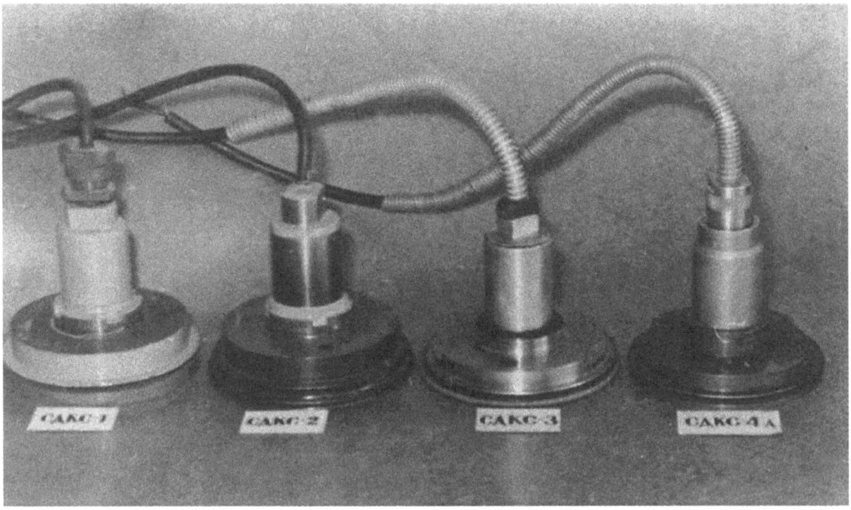

Figure 4-7. Several modifications of contact soil pressure cell SDKS depicted in Figure 4-6. General view.

did not exceed 0.028 MPa, and the mean deviation was approximately equal to 0.009 MPa, which is less than 2% of the maximum stress being measured.

A series of tests were carried out to determine the effects of temperature change on the performance of the cells under discussion. These tests were conducted on 15 instruments for different magnitudes in temperature change, e.g., 2, 3, 5, and 8°C. The average deviation in cell readings was 0.2 Hz per 1°C with a maximum deviation equal to 1.07 Hz per 1°C. In most practical cases, e.g., the investigation (monitoring) of soil/structure interaction such as retaining walls, footings, etc., this deviation is acceptable.

The original SDKS-1-0.5 contact soil pressure cell has been modified several times. This has resulted in better, more sensitive, more reliable, and 30% lighter instruments. One of the latest models, SDKS-6VMR (Figures 4-8 and 4-9), has an improved water sealing system, better design of the upper wire clamp, better protection for the cable, a shorter (47.5 vs 50 mm) sound wire, and many other improvements that have made this cell easy to assemble and install (Lazebnik 1982). It should be pointed out that the slight decrease in wire length and some increase in the allowable stress in the wire material resulted in a much higher (almost doubled) sensitivity of the instrument. Also, better air-tightness of the instrument's interior resulted in extended (minimum 10 years) performance in water-saturated soils, provided the instrument is not exposed to heavy dynamic loads and/or high temperatures (more than 50°C).

The pressure limitation of the SDKS-6VMR is 10 times higher than that

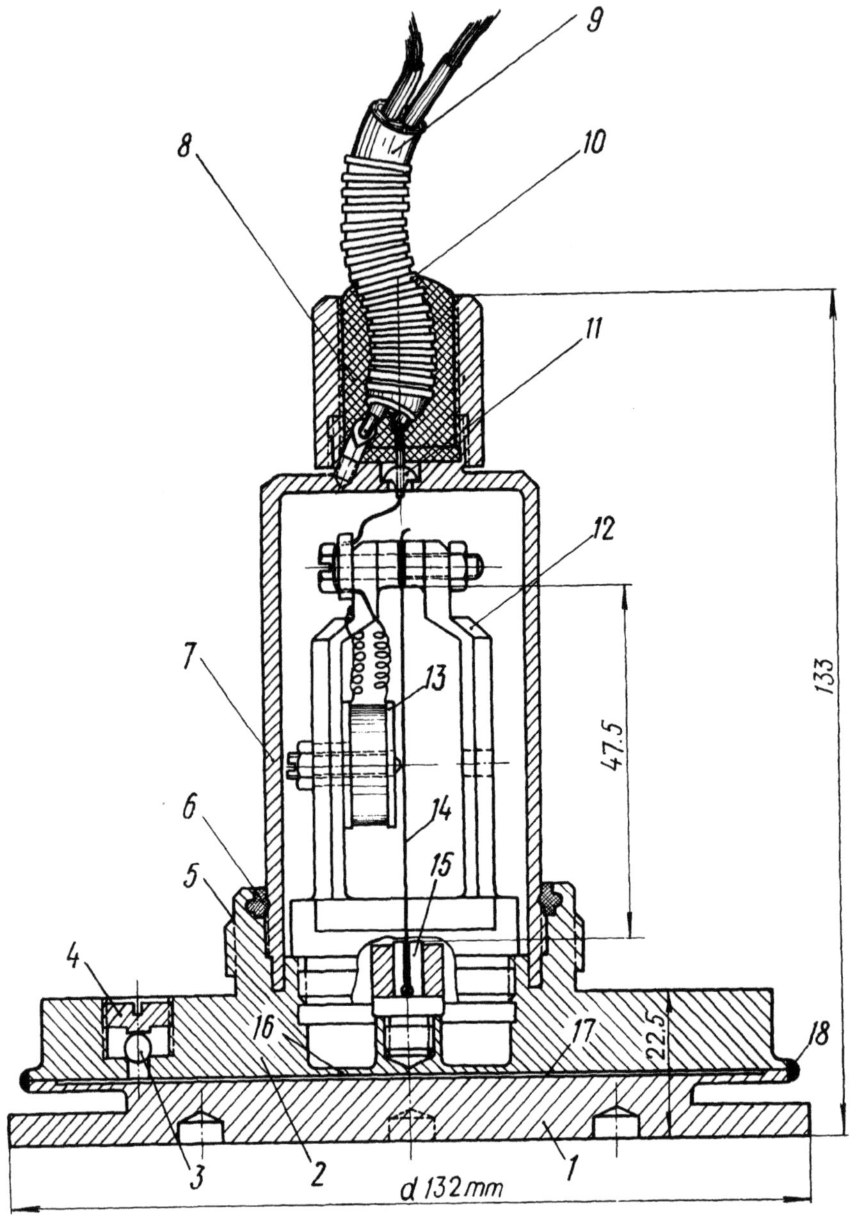

Figure 4-8. Contact soil pressure cell SDKS-6VMR. Typical cross section. 1. Soil pressure-bearing membrane; 2. body; 3. ball type; 4. nut; 5. thread; 6. seal; 7. protection cylinder; 8. sealing epoxy cement compound; 9. double-wire cable; 10. protection flexible metal hose; 11. insulator; 12. wire upper holder; 13. electromagnet; 14. wire; 15. wire lower holder; 16. measuring membrane; 17. hydraulic cavity.

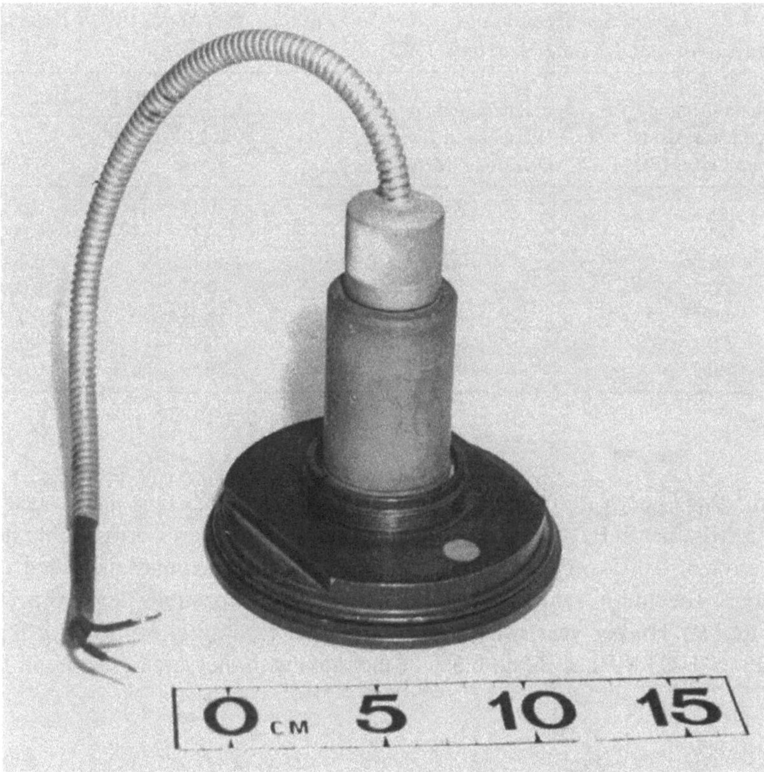

Figure 4-9. General view of contact soil pressure cell SDKS-6VMR.

specified for the SDKS-1-0.5. The SDKS-6VMR specifications are given in Table 4-2. The following are special technical characteristics of pressure cell SDKS-6VMR and relative errors of the instrument in percents.

Hysteresis: ≤ 1

Instability at cyclic load: < 0.7

Relative error in calibration by means of air pressure via layer of soil (sand): ≤ 1.5

Fluctuation of readings in several similar loadings via layer of soil: ≤ 2.8

Permissible overload: 10

Long-term laboratory tests and field observations conducted during a 10-year period have shown that the maximum drift in these cell readings could vary from

Table 4-2. Typical Specification of Contact Soil Pressure Cell SDKS-6VMR Depicted in Figures 4-8 and 4-9 (after Lazebnik 1982)

Maximum (Limiting) Pressure Measured by Pressure Cell (MPa)	Rated Modulus of Elasticity of Pressure Cell Body (MPa × 10^4)	Sensitivity per 1 Hz	
		% of Measurement Range	MPa
0.1	0.11	0.14	0.00014
0.3	0.32	0.13	0.00040
0.5	0.73	0.18	0.00090
1.0	1.60	0.22	0.00220
2.0	4.25	0.34	0.00670
3.0	6.42	0.40	0.01200
5.0	10.33	0.50	0.02500

3.3 to 5.9 Hz per year. The typical sensitivity of the SDKS-type instruments is 0.0015 MPa per 1 Hz, thus the aforementioned magnitude of drift results in a total error of 0.005 to 0.009 MPa or 1 to 1.8% of the maximum measured soil pressure. The mean values of drift in the readings constitute approximately 1.65 to 1.95 Hz per year, which results in total reading errors varying from 0.0025 to 0.003 MPa or 0.5 to 0.6% of the maximum measured pressure of the soil.

4.3 Contact Pressure Cell with Vibrating Wire Type Transducer for Mass Production

The contact pressure cells described in the previous section are relatively difficult to manufacture. They have been manufactured and assembled under laboratory conditions and used for some selected research projects. With an increased demand for these instruments, production under laboratory conditions has become increasingly difficult. Their relative complexity of these instruments was an obstacle to their efficient mass production. To overcome these problems, a new modification of this instrument, SDKS-7, has been introduced. Lazebnik et al. (1989) Its typical cross section and general view are shown in Figures 4-10 and 4-11. This instrument has been manufactured and supplied in sufficient numbers as required by the former Ministry of Energy and Electrification, USSR. This instrument consists of two basic parts: the "pressure receiving" body and the vibrating wire type transducer. The pressure receiving body consists of a hydraulic multiplicator and cell body, both similar to that used in the original contact pressure cell design. The transducer is screwed into the body of the cell, and the pressure measuring membrane, which is part of the transducer, is lined up with the hydraulic cavity. Next, an epoxy-based seal is installed, and the hydraulic

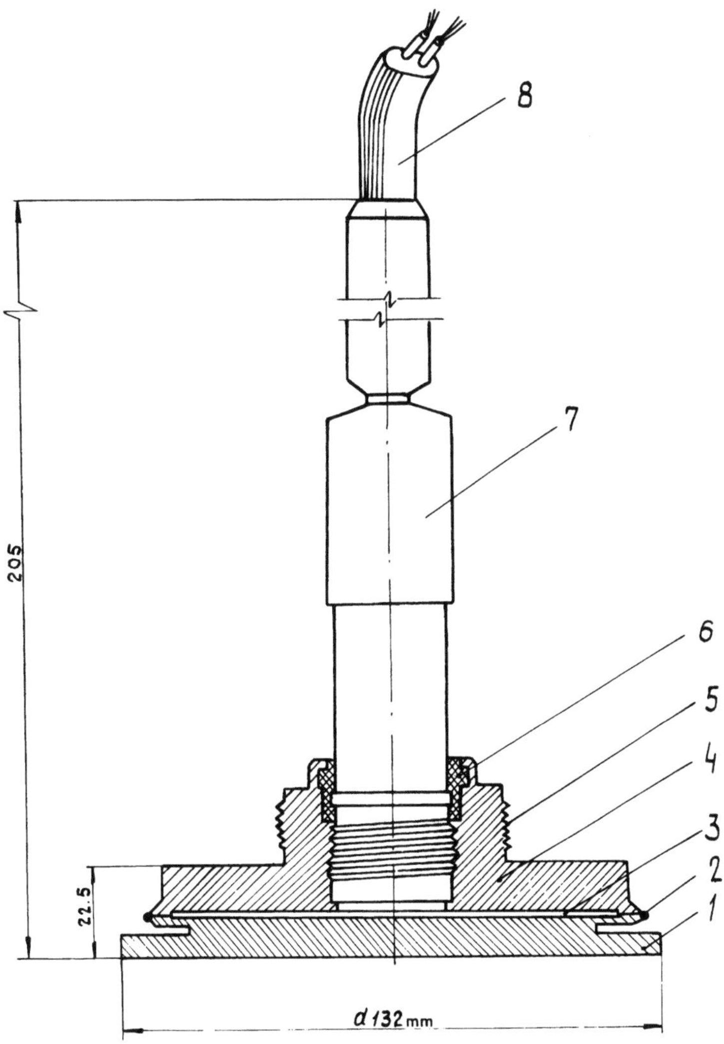

Figure 4-10. Contact soil pressure cell SDKS-7. Typical cross section. 1. Soil pressure-bearing membrane; 2. circumferential weld; 3. hydraulic cavity; 4. body; 5. thread; 6. sealing epoxy-based compound; 7. vibrating wire type transducer; 8. double-wire cable.

cavity is filled with a special liquid, similarly to the earlier modifications of the SDKS.

Representative calibration curves of an SDKS-7 are illustrated in Figure 4-12. As is seen from this figure, the cell was calibrated by air pressure and by noncompacted and compacted soil (medium sand). All three curves are practically linear. The maximum difference in calibration by air and compacted soil consti-

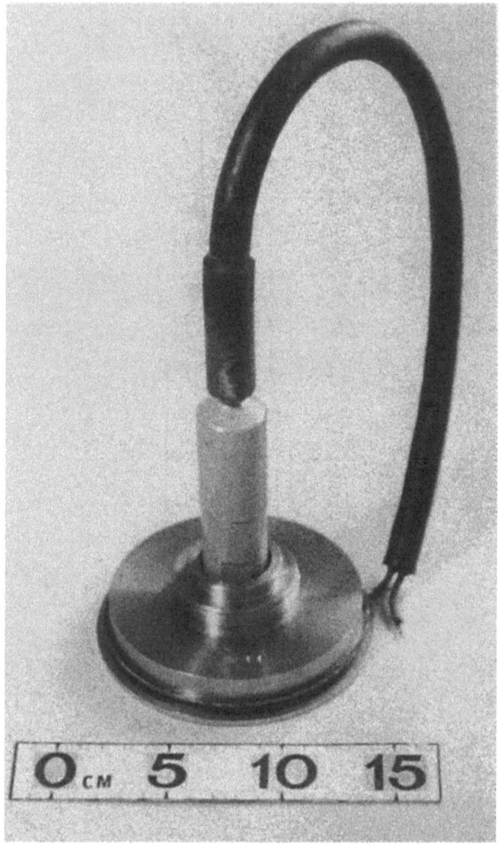

Figure 4-11. Contact soil pressure cell SDKS-7. General view.

tutes approximately 2.3%, and the difference between calibration results in non-compacted and compact soil is only 0.6%.

Other characteristics of the SDKS-7, e.g., hysteresis, instability during cyclic loading, and deviation of readings in several identical tests are very close to those characteristic for the SDKS-6VMR, as described in Section 4.2. The performance of three SDKS-7 cells has been extensively investigated in the laboratory, with readings taken for more than 7 months. This investigation has confirmed that the SDKS-7 cell is a reliable, robust, and sensitive instrument suitable for long-term performance in the most unfavorable conditions, for example, in the presence of soils having a high moisture content. SDKS-7 cells with increased (doubled and tripled) areas of soil contact membrane have also been tested. As expected, these tests have confirmed that larger contact areas result in increased sensitivity of the instrument and more accurate measurement of soil pressures.

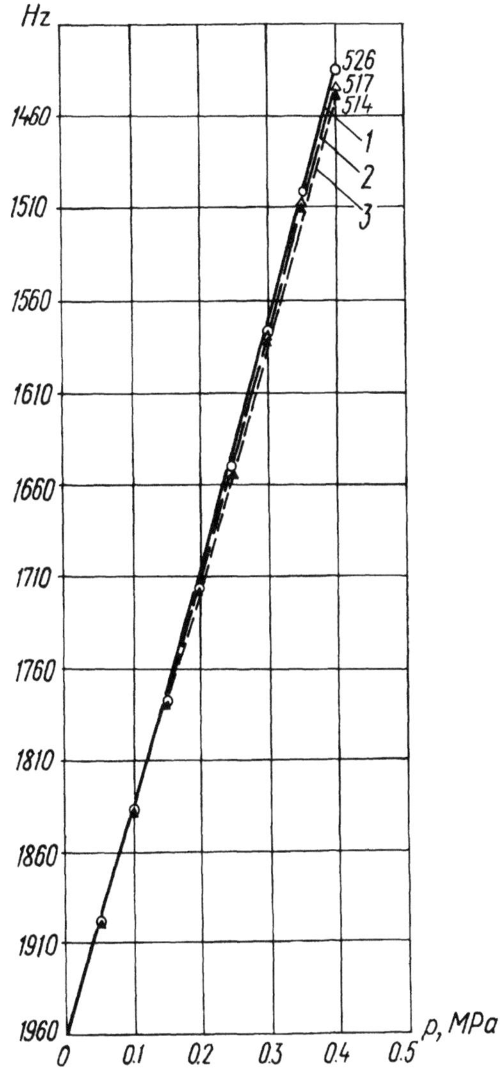

Figure 4-12. Calibration curves of contact soil pressure cell SDKS-7. 1. By air pressure; 2. by loose sand; 3. by compacted sand.

4.4 Effects of Cell Size on Scatter in Results of Soil Pressure Measurement

In practical construction projects, the foundation soil or backfill materials usually are not uniform in composition or density. In most cases properties of these soils such as granular composition and compressibility vary from section to section.

Soils may also contain some hard inclusions or have cavities. Therefore, the soil pressure measured by the soil pressure cell represents the soil/structure interaction condition only at the particular area of the structure, and not necessarily the averaged pressure. Furthermore, the instrument readings may also be affected by structure movements, for example, away from or toward the backfill soil.

Common sense suggests that the larger the contact area between the instrument and soil, the more accurate the measurement of soil pressure. Apparently with this in mind, the Japanese company Kyowa (1973) manufactured a soil pressure cell with a rather large diameter equal to 160 mm and 600 mm. Muhs (1947) reported results of soil pressure measurements against a 24-m-high wall that was backfilled with loose, clean sand in a manner that provided for a relatively uniform soil distribution behind the wall. This wall was part of a massive reinforced-concrete underground structure and was considered to be practically unyielding. The lateral pressures against this wall were measured by means of large (305-mm diameter) vibrating wire type cells. The area of the cell in contact with the soil was increased by placing specially designed metal plates between the soil and two of the cells. The outer surface of each plate was flush mounted at the surface of the wall in contact with the soil. In this setup, pressures against 54% of the height of the wall were measured over a width of 61 cm. This resulted in a consistent, smooth sequence of measured pressures that could be obtained by the use of small individual cells.

A representative example of the soil pressure diagram measured by Muhs (1947) is depicted in Figure 4-13. Quite different results were reported by Goldbeck (in Tschebotarioff 1958, 1973). Goldbeck investigated soil pressure against a massive gravity wall (bridge abutment) apparently backfilled with a cohesive soil. Soil pressures were measured by means of pressure cells 140 mm in diameter. The representative soil pressure diagram is depicted in Figure 4-14. Each point in this diagram is an average measurement from two instruments installed at the same level. The inconsistency of this diagram apparently stems from the nonuniformity of the backfill material because in practice it is very difficult to place cohesive type soils with uniform distribution and density. Under such conditions it is very difficult to obtain a smooth sequence of measured pressures by using relatively small individual cells.

Theoretical relationships between the nonhomogeneity of the material and the scatter in the instrument readings have been proposed by Lomakin and Sheinin (1970), Kheifitz (1973), Sammal (1980), and others. It should be noted that normally to solve the problem theoretically, knowledge of some specific soil and instrument parameters is required. These parameters, however, in most practical cases are not available to the researcher, which often makes the theoretical approach impractical.

In 1962 to 1964 this writer and co-workers carried out a number of tests in granular soils (sand) to determine experimentally variation in value of coefficient of soil lateral pressure "at rest" K_0 and the effects of soil nonuniformity (presence

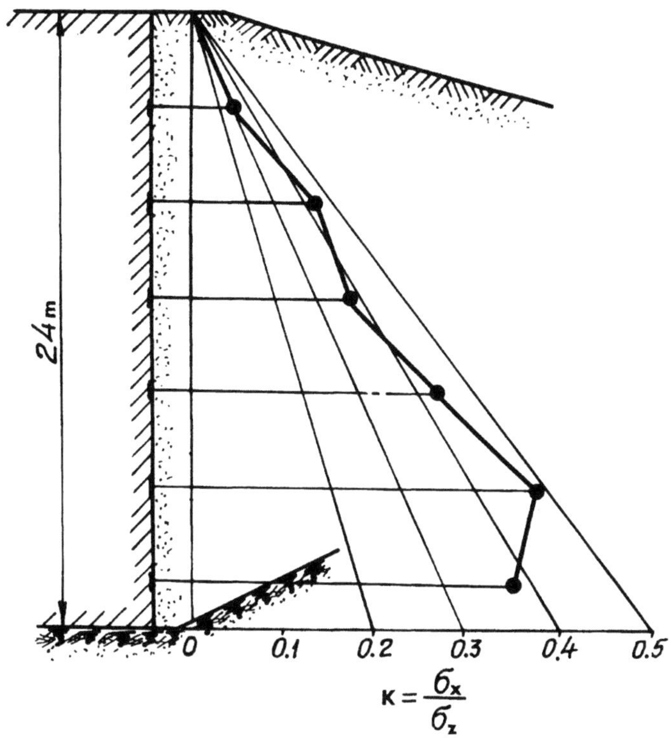

Figure 4-13. Lateral soil pressure of soil on a gravity-type retaining wall obtained from field measurements (after Muhs 1947).

of particles of different sizes) on scatter in soil pressure readings. The investigation was carried out in calibration chambers SCC described in Chapter 3. Sand was carefully placed in the chamber with no possibility of lateral expansion, hence the "at rest" condition was created. The pressure of the soil was measured by means of stiff flat cells, with the diameter of the soil pressure-bearing membrane equal to 120 mm. The tests were carried out in loose and densified sands. The test results are illustrated in Figure 4-15.

Each experimental point presented in this figure was obtained by statistical processing the data from eight tests. Despite extremely careful experimental procedures, the test points were quite scattered; they were more scattered in the tests conducted in loose sand than in compacted sand. This can be explained by the fact that contact between the instrument and the soil is more intimate in compacted sands than in loose sands. From a great number of laboratory experiments and those carried out on large-scale models tested under field conditions, it was found that the diameter of the cell plays a significant role in scatter of cell readings. As one can expect, the scatter of soil pressure measurements is

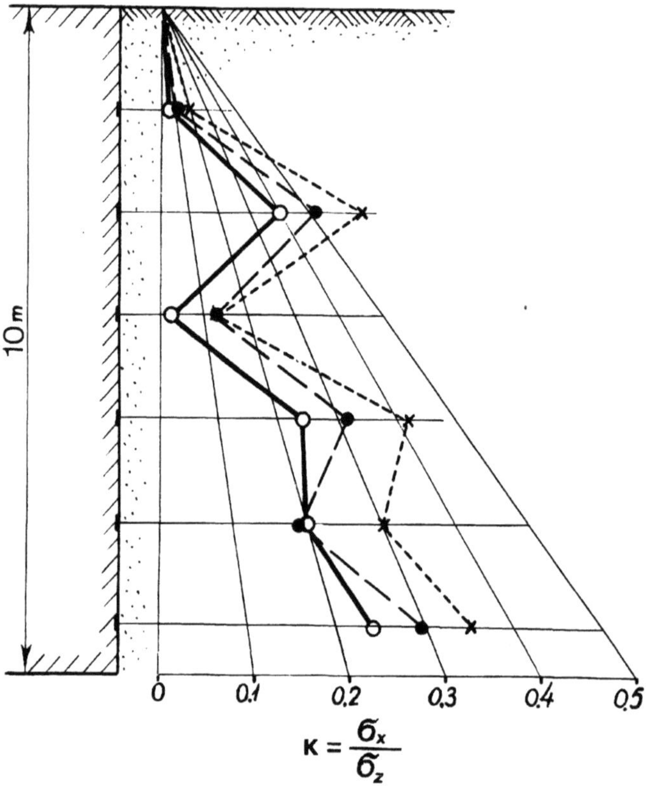

Figure 4-14. Experimental diagrams of lateral pressure of soil against bridge abutment.

more significant when measured by an instrument having a smaller-diameter contact surface. It must be pointed out that not only the diameter of the instrument but also the way the soil pressure-bearing membrane deforms, e.g., deflecting at the middle part or moving parallel from its original position, affects the scatter (error) in soil pressure readings. This is discussed in detail in the following chapters.

Statistically processed data from the aforementioned experiments are presented in Figure 4-16. In this figure p_m is the soil pressure measured by the particular instrument, and p is the mean value of measurements taken by instruments working under identical test conditions. During processing of the data included in Figure 4-16, the soil pressure measurement errors by some instruments, attributed to nonparallel movements of the soil pressure-bearing membrane, have been excluded using calibration data.

In the former USSR the permissible limit of scatter of soil pressure measurements by SDKS-type pressure cells (details on these cells are provided earlier in this chapter) is regulated by *Guidelines for Determination of Contact Soil*

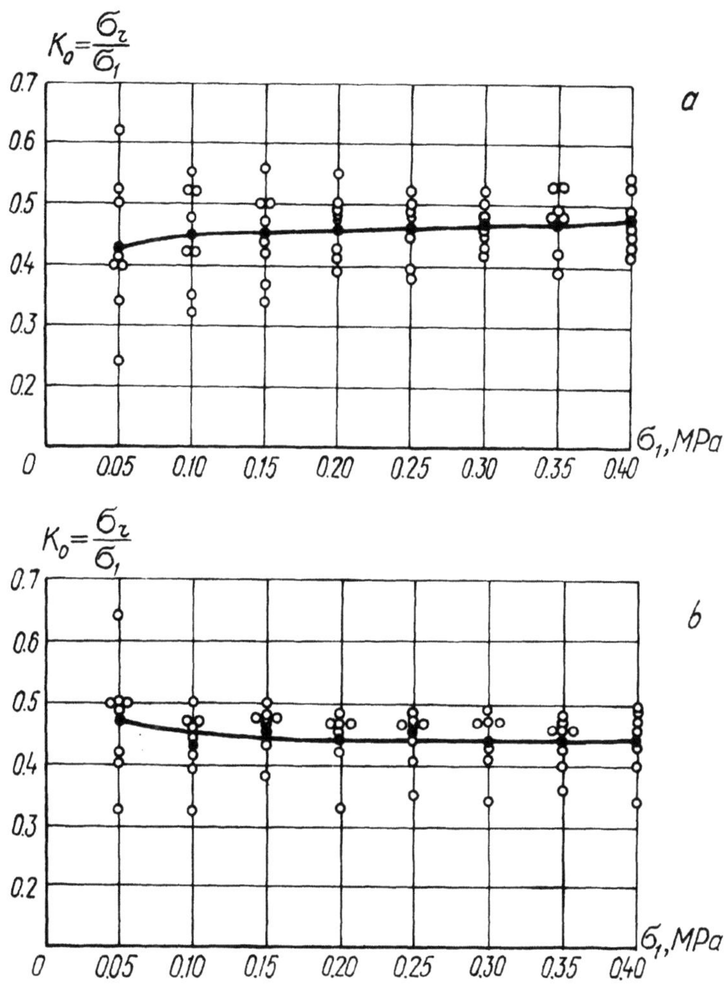

Figure 4-15. Variation in coefficient of lateral soil pressure K_0 with increase in vertical pressure. (a) Loose sand; (b) compacted sand.

Pressure Using Pressure Cells SDKS (1967). According to this document, the scatter of soil pressure measurements must not exceed 15 to 20% of the p_m/p value. As is seen from Figure 4-16, the value of p_m/p for instruments with d = 50 mm jumps up to 80%, for d = 100 mm it is within 20 to 30%, and beginning from d = 120 mm the value of p_m/p is approximately within a permissible 15% or less.

From these results it is quite obvious that in order to reduce the error in measurements of soil pressure caused by soil nonhomogeneity, the size of the soil contact membrane should be as large as possible, especially when it is used

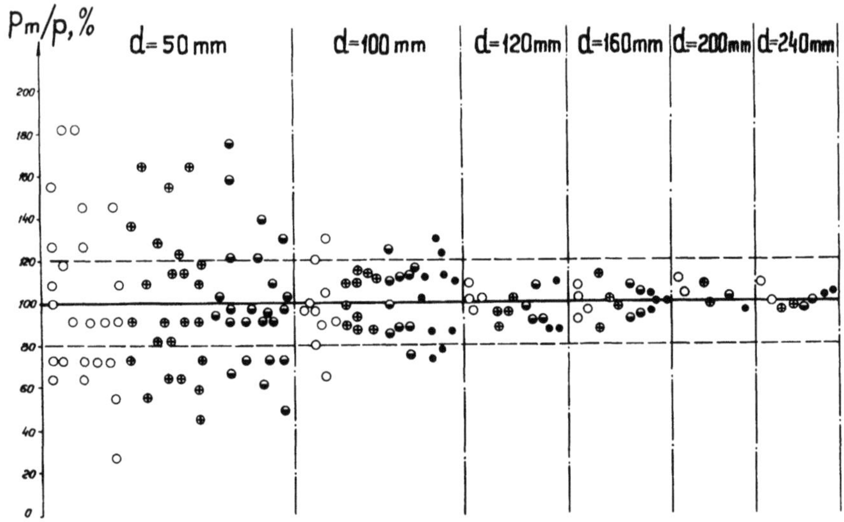

Figure 4-16. Scatter of cell readings as a function of diameter of soil pressure-bearing membrane.

for measuring relatively small intensity pressures. For better confidence in the obtained results the process of stress measurement should be at least duplicated. This is particularly true for field tests in which repetition of the experiments is not possible.

In conclusion, we state that the best result in the measurement of soil pressure can be achieved when pressure cells are installed (mounted) in groups of three to four or covered by rigid plate flush-mounted with the underside of footings or the backside of the retaining structure. This averages out the soil reaction or lateral pressure and local discontinuities in the soil over the larger contact area of the plate. For small-diameter individual cells these discontinuities can become very pronounced and obscure the overall picture.

5

Errors in Measuring Soil Contact Pressures by Means of Nonstiff Pressure Cells

5.1 Introduction

In this work nonstiff soil pressure measuring cells are defined as those that are less stiff than concrete, and the "error" in the measuring of soil pressures is defined as the deviation between the measured and the true value of the soil pressure. Thus, mathematically the error can also be interpreted as the accuracy of the soil pressure measurements. In previous chapters it has been shown that the stiffness of the soil pressure measuring cells greatly affects the instrument's reading. It has also been illustrated that when the instrument stiffness is close or equal to that of concrete (e.g., soil pressure cells with hydraulic multiplicators) the error in soil pressure measurement is minimal, provided the founding or backfill soils are homogeneous and uniform.

As discussed earlier, two basic types of earth pressure measuring cells are currently used: membrane-type cells and cells with a hydraulic multiplicator. Membrane-type cells, which are composed of a stiff circular membrane fully supported on a stiff edge ring, are often used by investigators for measurements of soil pressures. The membrane's deflection is registered by either an electrical resistance strain gage bonded directly on the interior face of the load-bearing membrane, or by a vibrating wire transducer, as discussed earlier. Performance evaluations of various types of commercially available earth pressure measuring cells and recommendations for cell design and manufacture are given in Brown (1977), Corbett et al. (1971), Eidelman (1960b), Hvorslev (1976), Lazebnik (1969, 1980), Lipovetskaya (1953), O'Rourke (1978), Reese et al. (1968), Selig (1964), State of California (1968, 1971), Uginchus and Bombchinsky (1954), and Weiler and Kulhawy (1978, 1982).

Unfortunately, published information on soil pressure cells is usually very sketchy and in most practical cases does not provide sufficient data on potential errors in soil pressure measurements. In the mid-1960s this author, with co-workers, carried out a study aimed at investigating the effects of cell stiffness (working membrane displacement) and changes in soil parameters under compressive loads on the accuracy (errors) of instrument performance.

In the investigation process other factors that affect soil pressure measurement by nonstiff cells, e.g., proximity of instruments to each other, depth of layer of soil above the active membrane of the instruments, and others have been investigated. These are discussed further in the following sections.

5.2 Instruments and Devices Used in Experiments for Determining the Errors in Soil Pressure Measurements by Nonstiff Cells

Three basic types of nonstiff pressure cells were used in these studies. The movements (deflection) of these cells' active surfaces can be described as parallel displacement, flexural deformation (strain), or a combination of both. Basic structural sketches of these cells are depicted in Figure 5-1. The stiffness of these instruments was similar to that of instruments frequently used in the past in different investigations of the phenomenon of soil/structure interaction. Basic technical parameters of the experimental test cells are given in Table 5-1.

Furthermore, in order to investigate the possible connection between errors in the measured pressure and diameter of the active surface of the pressure cell, the test cell membranes were manufactured with the same stiffness ($E_{epc} \approx 210$ MPa) but with soil pressure-bearing membrane diameters varying from 20 to 80 mm. Also, a large-diameter soil pressure measurement device (an idea for which was proposed by professor M. N. Goldstein of Dnepropetrovsk, Ukraine, Civil Engineering Institute) was used to determine the total soil pressure applied to a large area, as well as the distribution of this pressure over the pressure-bearing membrane. A general view of this device is shown in Figure 5-2.

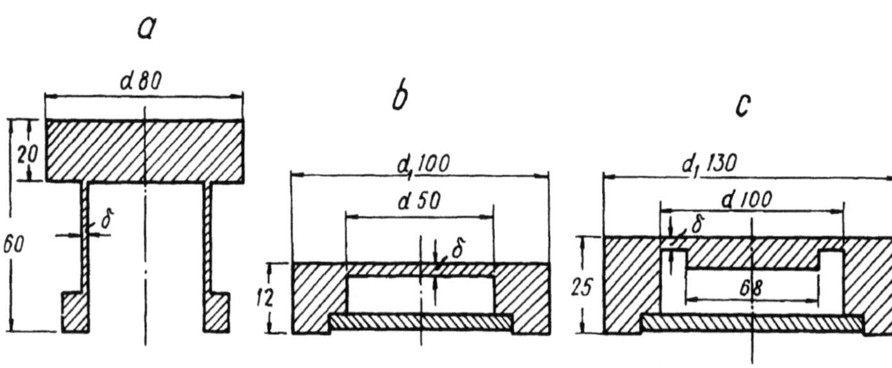

Figure 5-1. Soil pressure cells used in investigations of systematic error in soil pressure measurements by nonstiff instruments. (a) Mushroom type; (b) and (c) membrane type.

Table 5-1. Parameters of Pressure Cells Used in Investigations for Determination of Systematic Error in Measurements of Soil Pressures

Pressure Cell	Pressure Cell No.	Thickness of Wall or Membrane, δ (mm)	Displacement at Center of Contact Surface (Membrane) at Pressure 0.1 MPa (μm)	Modulus of Elasticity of Pressure Cell (Effective or Nominal) (MPa)
Depicted in Figure 5-1a	1	0.23	2.48	2,410
(d = 80 mm)	2	0.30	1.86	3,420
	6	0.90	0.97	6,845
	8	1.23	0.47	13,700
Depicted in Figure 5-1b	2	0.90	43	28
(d = 50 mm)	3	1.0	31.25	38.5
	3′	1.3	14.25	84
	4	1.5	9.25	130
	5	1.75	5.75	209
	6	2.0	4.0	299
	7	2.3	2.75	435
Depicted in Figure 5-1c	1	1.6	16.6	152
(d = 100 mm)	2	1.8	14.2	196
	3	2.04	9.52	267
	4	2.22	7.25	371
	5	2.63	4.75	527
	6	3.01	3.75	667
	7	3.35	3.13	800

Figure 5-2. Multicell large-diameter (d = 400 mm) soil pressure measurement device.

The device is comprised of a large-diameter (d_l = 550 mm; and d = 400 mm) steel membrane 15 mm thick with a number of small pressure cells installed within three radial slots that are milled on the membrane's surface. The active surfaces of the small cells are flush with the membrane of the large soil pressure measuring device. After installation of the individual cells, the space between adjacent instruments is sealed. The total value of the soil pressure is measured by means of standard bonded toil resistance strain gages mounted on the inner surface of the large membrane. The modulus of elasticity of this device is 72.5 MPa. Deformation of walls and/or membranes of the pressure cells also has been measured by means of standard toil resistance strain gages.

The investigations were carried out in calibration devices SCC and LCC, which were described in Chapter 3. Displacement (deflection) of the contact surfaces of the pressure cell membrane under load were determined analytically as well as by means of mechanical dial gages whose accuracy is 0.001 mm. Depending on the consistency of the test results, all tests were repeated 3 to 5 times. The typical number of loading cycles was two, and test loads were applied in increments of 0.05 and 0.1 MPa with a maximum pressure reached of 0.4 MPa. Readings at the first and second unloading cycles were also obtained. For comparative purposes, the test cells were also loaded by air, applied via a rubber diaphragm (no soil present). Results of these experiments were taken as 100% ("error free") and were considered the effective applied pressure p. Determination of the systematic measurement error, Δ_p, is discussed in the following sections.

5.3 Changes in Properties of Cohesionless Soils That Occur Due to Effects of Compression

Primarily medium- and coarse-grained river sands (a cohesionless soil) were used in the experiments. Grain size distribution and other information on these sands are given in Table 5-2. Cohesionless soils were used because the maximum error in the soil pressure measurement, caused by the difference in stiffness characteristics between the pressure cell and the soil (when measuring pressures in a mass of soil), or between the soil pressure cell and the material of the structure in which this cell is installed, manifests itself in soils having considerable internal friction (sands resist distortional displacement essentially by internal friction).

The shear characteristics of granular soils that were used in the experiments and the changes of these soils' angle of internal friction, ϕ, were investigated in a specially designed direct shear test machine. Vertical stresses in these tests were applied at increments of 0.1 MPa with the maximum vertical stress equal to 0.4 MPa. The results of these tests are given in Table 5-3.

Additionally, changes in sand parameters, such as the modulus of deformation and coefficient of lateral pressure, under applied pressures were investigated in a special apparatus that is shown in Figures 5-3 and 5-4. In this investigation the sand was confined in large steel cylinders (rings) 260 and 300 mm in diameter and loaded by means of a hydraulic press. The wall thickness of these cylinders and their depth were, respectively, equal to 6.4 and 6.5 mm and 130 and 100 mm.

The cylinder interiors were of thoroughly polished chrome-plated construction to achieve a mirror-state condition. Cylinders were installed on a very stiff steel base plate, also with a polished chrome-plated surface. Test soils were placed into the cylinders in thin horizontal layers and compacted where required to achieve a desirable dense condition. Compaction was carried out in layers, each of which was compacted by an equal number of blows produced by a massive cast iron cylindrical block that was dropped from a constant height.

The vertical stress produced by the precise mechanical press was transmitted to the carefully leveled surface of sand installed in the cylinder via a very stiff steel bearing cover whose diameter was slightly smaller than the interior diameter of the cylinder. The underside (bearing) surface of the cover was also chrome plated and polished to the same mirror-state condition. Vertical displacements of the bearing cover under an applied load were measured by standard dial gages with a relative accuracy of 0.01 mm. Stresses in the cylinders, associated with vertical stresses applied to the sand placed inside of these cylinders, were measured by means of eight pairs of standard toil resistance strain gages posted on the outside surface of each cylinder. A detailed description of this apparatus is given in Lazebnik (1967).

The load on the cylinder's cover plate was applied in increments of 0.1 MPa, and each successive step of loading was performed after stabilization of soil

Table 5-2. Grain Size Distribution and Other Characteristics of Soils Used in Experiments Described in Chapter 5

	Quantity of Particles (mm), Percent of Weight					Density of Dry Soil ρ_d (kN/m^3)		Void Ratio (e)		Angle of Response (deg)	
Soil	3–2	2–1	1–0.5	0.5–0.25	0.25–0.05	Loose	Dense	Max.	Min.	Loose	Dense
Medium Sand	0.4	0.55	8.2	49.95	44.9	15.8	18.2	0.68	0.45	33	36.5
Coarse Sand	—	4.5	85.1	7.5	2.9	16.4	17.9	0.603	0.469	33.5	37

Table 5-3. Changes in Shearing Strength and Values of Angle of Internal Friction of Sands as Related to Changes in Vertical Stresses

Type of Sand		Vertical Pressure (MPa)	Shearing Stress (MPa)	Friction Coefficient	Angle of Internal Friction (deg)
Coarse Sand	Loose $D_r < 30\%$	0.1	0.0642	0.642	32°40′
		0.2	0.1217	0.609	31°20′
		0.3	0.1875	0.625	32°00′
		0.4	0.2358	0.590	30°30′
	Dense $D_r > 67\%$	0.1	0.0933	0.933	43°00′
		0.2	0.1833	0.917	42°30′
		0.3	0.2750	0.917	42°30′
		0.4	0.3612	0.903	42°05′
Medium Sand	Loose $D_r < 30\%$	0.1	0.0481	0.481	25°40′
		0.2	0.0992	0.496	26°20′
		0.3	0.1500	0.500	26°35′
		0.4	0.1942	0.486	25°55′
	Dense $D_r > 67\%$	0.1	0.0808	0.808	38°15′
		0.2	0.1567	0.784	37°25′
		0.3	0.2350	0.783	36°15′
		0.4	0.2967	0.742	36°10′

Note: D_r = soil relative density.

Figure 5-3. Soil testing apparatus. General view.

settlement. The loading was carried out in two cycles. During the first cycle, sand was loaded in increments as indicated in Table 5-4 (from 0.05 to 0.4 MPa) and then removed in similar increments. During the second cycle, the same type of load was applied on the sand already compressed to some degree during the first stage of loading. The value of lateral soil pressure inside the cylinder was determined analytically from measured tensile stresses in the cylinder wall caused by the known vertical load applied to the soil's surface. The modulus of total deformation of the soil, E_s, was calculated from the following relationship:

Figure 5-4. Basic components of soil testing apparatus shown in Figure 5-3.

$$E_s = \beta \frac{d\sigma_z}{d\varepsilon} \tag{5-1}$$

where $d\sigma_z$ is the increment of vertical stress and $d\varepsilon$ is the respective increment of relative deformation of soil in the cylinder. The value of the factor β is calculated from the following formulation:

$$\beta = \frac{1 - K_0}{1 + K_0}(1 + 2K_0) \tag{5-2}$$

where K_0 is the coefficient of soil lateral pressure "at rest." K_0 is an average value of numerous readings taken from strain gages installed on the cylinders' outside faces. For details consult Chapter 10.

Because sands, like the majority of other soils, do not deform linearly, the values of E_s and β obtained by equations (5-1) and (5-2) can be regarded as approximate. It will be shown later that these approximations are adequate for the described experimental conditions.

From the data presented in Table 5-4, it follows that the sand modulus of deformation, E_s, increases substantially with increases in load application in the first cycle of loading. However, a dramatic increase in the soil modulus of

Table 5-4. Experimentally Obtained Moduli of Deformation of Soils, E_s

Soil	State of Soil	Loading Cycle	Moduli of Deformation E_s at Pressure p (MPa)				
			0.05	0.10	0.20	0.30	0.40
Medium Sand	Loose	First	8.4	10.7	14.4	18.2	20.7
		Second	81.0	90.5	104.0	108.0	110.0
	Dense	First	40.0	82.6	94.9	121.0	147.0
		Second	109.0	136.0	223.0	234.0	234.0
Coarse Sand	Loose	First	18.2	32.2	52.5	85.5	107.0
		Second	88.0	128.0	163.0	211.0	226.0
	Dense	First	55.0	85.0	115.0	212.0	212.0
		Second	137.0	174.0	265.0	353.0	400.0

deformation takes place in the second loading cycle. This phenomenon takes place despite the fact that only insignificant increases in sand density under the static compressive loads occur. For example, in the investigation in question, after three consecutive static loadings that were applied in increments of 0.1 MPa and up to 0.6 MPa, the porosity as well as the density of a loose medium sand resulted in an increase in soil density of about 2%. At the same time the modulus of deformation of this soil had increased 16-fold.

5.4 Effect of Distance Between Adjacent Soil Pressure Cells on Instrument Performance

This effect of distance was tested in the calibration chambers described earlier in Chapter 3. The soil pressure cells depicted in Figures 5-1a and 5-1c were used in experiments that were carried out to determine the effects of distance between adjacent cells on their performance. For this purpose these cells were installed into special inserts that subsequently were installed into the calibration chamber and loaded there. The inserts, made from steel, concrete and acrylic material having a modulus of elasticity equal to 2000 MPa, were designed to accommodate one, two, and up to five pressure cells. One of these inserts made to accommodate five cells is shown in Figure 5-5.

Figure 5-5. Five-pressure-cell insert made from acrylic plastic. General view.

Readings obtained from the single-cell installation were taken as 100% (reference) and compared with readings from arrangements of two, three, and more cells. As usual, all tests were, as a minimum, duplicated. The investigation indicated that the effects of adjacent cells upon each other manifest themselves when the distance between the cells equals $1.5d_1$ or less; at a distance equal to $3d_1$ or more, there are no effects whatsoever. (d_1 is the diameter of the individual cells.)

Other special investigations indicated that the error in readings taken from pressure cells installed into "soft" acrylic inserts in comparison with concrete inserts is relatively small (approximately 2%). This effect is more pronounced where less stiff instruments are used. It can be explained by the fact that the concentration of soil pressure above the cell is more likely to occur when stiff contact pressure cells are installed into relatively soft acrylic inserts. Note that in the case in question the modulus of elasticity of the stiffer pressure cell (Figure 5-1a) was seven times that of the acrylic insert. The concentration in soil pressure is due to the fact that under the load the acrylic insert deforms more than the stiff cell does. This results in a slight projection of the cell's active surface above the insert's surface; the latter results in soil pressure concentrations above the instrument's membrane.

5.5 Effect of Depth of Soil Above Instrument's Pressure-Bearing Membrane on Instrument Performance

Insufficient soil layer depths (H_s) in the calibration chamber above the cell's pressure-bearing membrane can result in substantial errors in readings of soil pressures taken by the instrument.

According to Peattie and Sparrow (1954) and Trollope and Lee (1961), an insufficient depth of soil typically results in irregular stress distribution in the stressed layer of soil. According to these workers, this usually happens when the depth of the outlying layer of soil is less than $0.9d_1$, where d_1 is the diameter of the instrument's active membrane. Trollope and Lee (1961) concluded that the formation of shear zones and arches in the overlying soil are responsible for the aforementioned phenomenon.

Lazebnik and Smirnov (1972), on the basis of their experiments in which three different types of pressure cells were used, concluded that the $0.9d_1$ criterion for arch formation in soil between the adjacent instruments and other factors that affect soil pressure cell performance apply only in the case of a rather stiff pressure cell, or where soils with a small modulus of deformation are used. The depth criteria for less stiff instruments and greater values of the soil modulus of deformation can substantially exceed the $0.9d_1$ criterion. The depth of the overlying soil also depends on the character of deformation of a cell's active surface under the load.

Figure 5-6 presents diagrams illustrating the effect of the depth of the soil layer placed above the soil pressure-bearing membrane. It is given in the form of the p_m/p ratio (in percents) as a function of H_s/D_{ca} for the membrane-type cells depicted in Figure 5-1b. Characteristics of these cells are given in Table 5-1. Here, p_m and p are, respectively, the measured soil pressure and the actual stress that may exist at the level of the soil pressure membrane in absence of the cell, and H_s and D_{ca} are respectively the depth of the soil layer above the cell's active surface and the diameter of the calibration device SCC. The H_s/D_{ca} ratio is defined as the relative thickness of the layer of soil above the soil pressure-bearing membrane. The experiments were conducted in the calibration chamber SCC-475-0.5 described earlier in Chapter 3; medium loose and dense sands were used in this investigation.

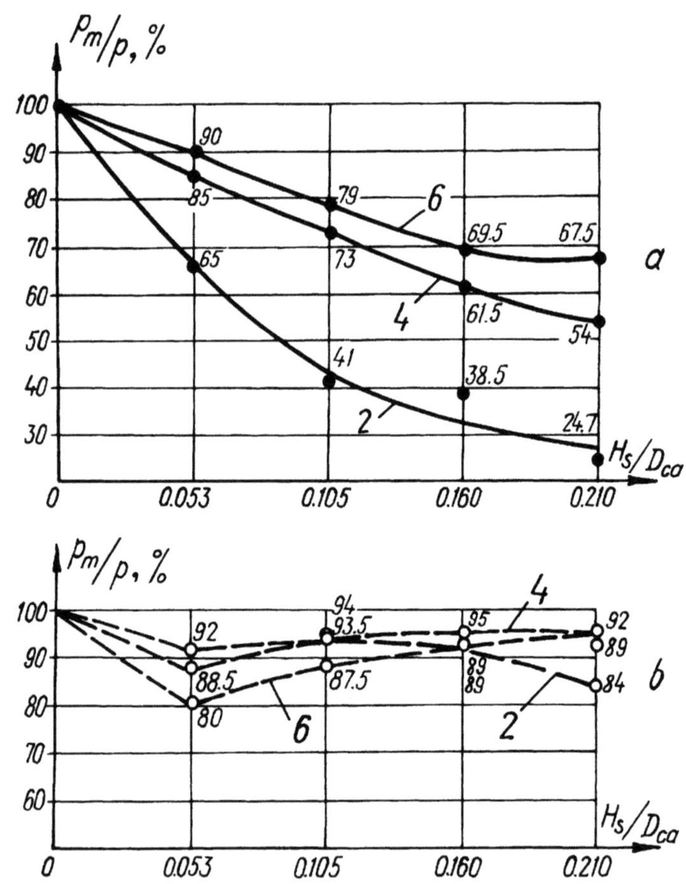

Figure 5-6. Effects of relative thickness of layer of soil (H_s/D_{ca}) on soil pressure reading. (Stiffness characteristics of cells No. 2 and 6 are given in Table 5-1.) (a) In dense sand; (b) in loose sand.

As is seen from Figure 5-6a, in dense sands the soil pressure measurement by cell No. 6 was practically stabilized at $H_s/D_{ca} = 0.16$, which corresponded to $H_s = 75$ mm, or 1.5 D_{ca}. Stabilization of readings of instruments No. 4 and No. 2 were achieved at, respectively, $H_s/D_{ca} = 0.263$ ($H_s = 125$ mm) and $H_s/D_{ca} = 0.315$ ($H_s = 150$ mm).

In experiments in loose soil (Figure 5-6b), stabilization in soil pressure readings by all instruments was achieved at approximately $H_s/D_{ca} = 0.105$ ($H_s = 50$ mm). The results of investigations conducted on the membrane-type pressure cells depicted in Figure 5-1c are illustrated in Figure 5-7. This investigation was carried out on two cells in both loose and dense medium sands. From curves $p_m/p = f(H_s/D_{ca})$ it is shown that cell readings are stabilized at H_s/D_{ca} equal to 0.21 or less. Under test conditions this corresponds to $H_s = 100$ mm.

From the data presented in Table 5-1, one can see that at a pressure equal to 0.1 MPa the pressure-bearing membrane of cells No. 4 and 6 (Figure 5-1b) deflects to 9.25 and 4.0 micrometers, respectively, and the deflection of the membranes of pressure cells No. 1 and 4 (Figure 5-1c) is, respectively, 16.6 and 7.25 mm. Despite the fact that the center of the contact membrane in cell No. 1 deflects more than the membrane of cell No. 4, stabilization of the measured pressures takes place in both loose and dense sands at a H_s/D_{ca} ratio less than unity; e.g., in cell No. 4 it takes place where the ratio H_s/D_{ca} approximately equals 0.9. The H_s/D_{ca} ratio in membrane-type pressure cell No. 6 (the stiffest among those used in the experiments) was approximately equal to 1.5.

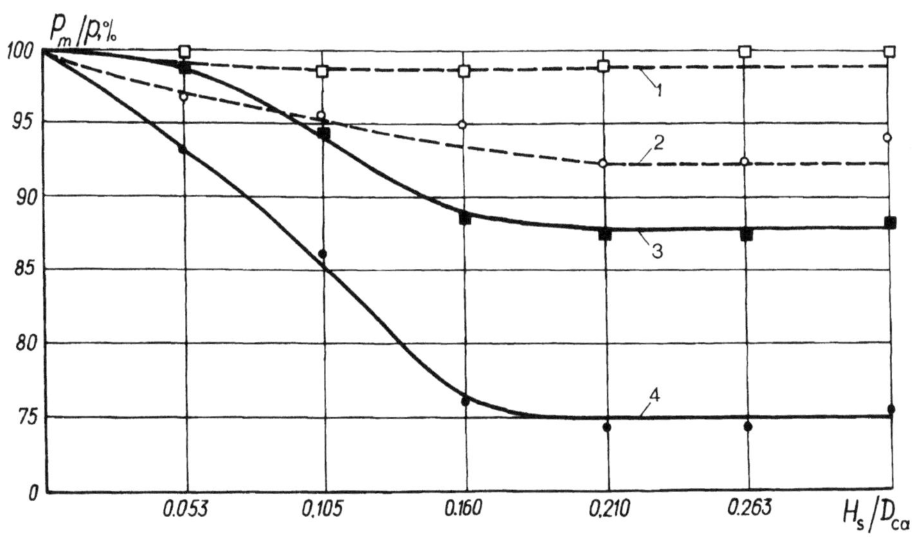

Figure 5-7. Diagrams $p_m/p = f(H_s/D_{ca})$ for membrane-type cell depicted in Figure 5-1c. 1 and 3. Cell No. 4 tested in loose and dense sand, respectively; 2 and 4. Cell No. 1 tested in loose and dense sand, respectively.

The effects of soil depth on cell performance during its calibration by pressure transferred to the soil-bearing surface via the layer of soil can also be found in Krivorotov (1971).

5.6 Errors in Soil Pressure Measurements by Nonstiff Soil Pressure Cells

As pointed out earlier, for a membrane-type cell it is important to minimize the effect of membrane deflection on stress concentration around the area near the rim of the cell; this is essentially the least active part of such a cell. On the other hand, the membrane must deflect so that a measurable strain is achieved. From this it follows that the deformation modulus at the center of the soil pressure-bearing membrane must be lower than that of the solid steel edge ring. The latter inevitably results in stress concentration on the edge ring. To diminish this effect on the soil pressure reading Peattie and Sparrow (1954) recommend that the ratio of the cell active (sensitive) area to the total facial area of the membrane-type cell be less than about 0.45. Uff (1970) suggested that, for diaphragm cells with vibrating wire transducers, this ratio be reduced to about 0.3.

This phenomenon has been investigated by Lazebnik and co-workers. In this investigation the diaphragm-type pressure cells shown in Figure 5-1c (having different stiffnesses) were used. All cells were tested in the calibration chamber installed in a concrete insert. Details on the calibration device are provided in Chapter 3.

From the previously conducted experiments it was found that membrane-type pressure cells installed in concrete inserts can produce credible results under the following conditions: the active pressure upon these cells does not exceed a value of 0.4 MPa, deflection of the soil pressure-bearing membrane is not more than 7.5 micrometers, and the soil modulus of deformation is maintained within a range of 10 to 400 MPa. Therefore, in the aforementioned experiments the soil pressure $p = 0.4$ MPa was taken as 100% (reference). All seven cells of the type depicted in Figure 5-1c (see also Table 5-1) were tested; for medium sands having different moduli of deformation the test results presented in graphical form are shown in Figure 5-8. The curves presented in this figure show the ratio p_m/p (%) as a function of the membrane deflection, f.

From the diagrams presented in Figure 5-8 it follows that despite the fact that the pressure cell stiffnesses were rather high (maximum deflection of soil pressure-bearing membrane at pressure $p = 0.4$ MPa did not exceed 0.07 mm), the relative measurement error Δ_p was in a range of 21–26% (curves 2 and 3).

Similar results have been obtained from testing the diaphragm pressure cells depicted in Figure 5-1b. In this case, however, the experimental points are substantially scattered, and the test results reveal a greater nonlinearity in comparison with the performance of the cells depicted in Figure 5-1c; this was particularly characteristic for cells having the least stiffness.

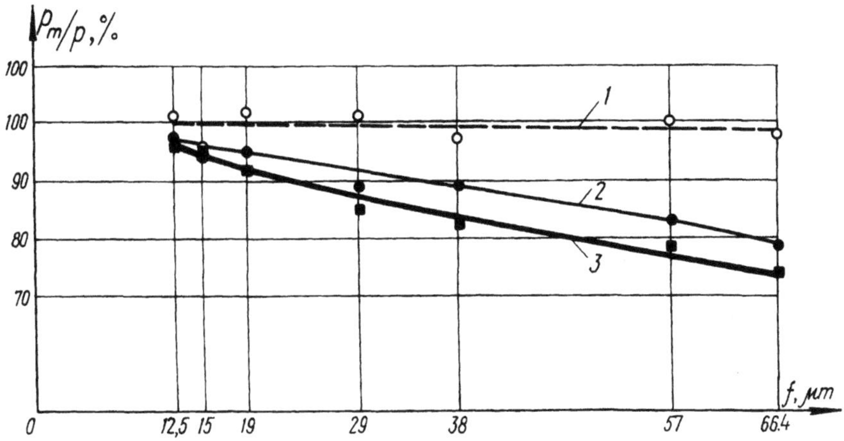

Figure 5-8. Relationship between p_m/p (%) and deflection of active surfaces of cells depicted in Figure 5-1c at different values of soil modulus of deformation. 1. At E_s = 20.7 MPa; 2. at E_s = 147 to 212 MPa; 3. at E_s = 234 to 400 MPa.

Baranov (1962) proposed the following equation for determination of the value of relative error of measurement of soil pressure, Δ_p, by contact pressure cells installed in a stiff structure:

$$\Delta_p = \frac{-1}{\frac{\pi(1-\mu_s^2)}{4} \times \frac{d}{f_{p=1}E_s} + 1} \tag{5-3}$$

where

$f_p = 1$ = maximum deflection (in cm) of soil pressure-bearing diaphragm at 0.1 MPa pressure

μ_s = Poisson ratio of test soil

d = diameter of soil pressure-bearing membrane

E_s = soil modulus of deformation.

Two basic shortcomings of this Baranov formula are as follows. First and foremost is the uncertainty in the determination of the value of $f_{p=1}$. Second, the formula assumes that the value Poisson's ratio of the soil, μ_s, is a constant value. However, similar to the soil modulus of deformation, E_s, μ_s can change when the soil is loaded. From this it was concluded that the systematic error of contact soil pressure measurement is basically related to the E_{epc}/E_s ratio, where E_{epc} is the modulus of deformation of the earth pressure cell. Accordingly, 680 tests on the cells depicted in Figure 5-1c have been carried out to determine the relationship between p_m/p (%) and E_{epc}/E_s. The results of these experiments are presented in Figure 5-9a (first loading cycle) and Figure 5-9b (second loading cycle). From curves that approximate the test results, one may conclude that an increase in E_{epc}/E_s values results in a rapid decrease in soil pressure measurements (p_m/p), and at $E_{epc}/E_s \approx 5$ the value of p_m/p does not exceed approximately 3%. Furthermore, an increase in relative cell stiffness ($E_{epc}/E_s \approx 10$) results in $p_m/p \approx 99\%$, or in other words the error in measurement in soil pressure contributes only approximately 1%.

Similar tests have been carried out using the membrane-type cells depicted in Figure 5-1b. The results of these tests are illustrated in Figure 5-10. As seen

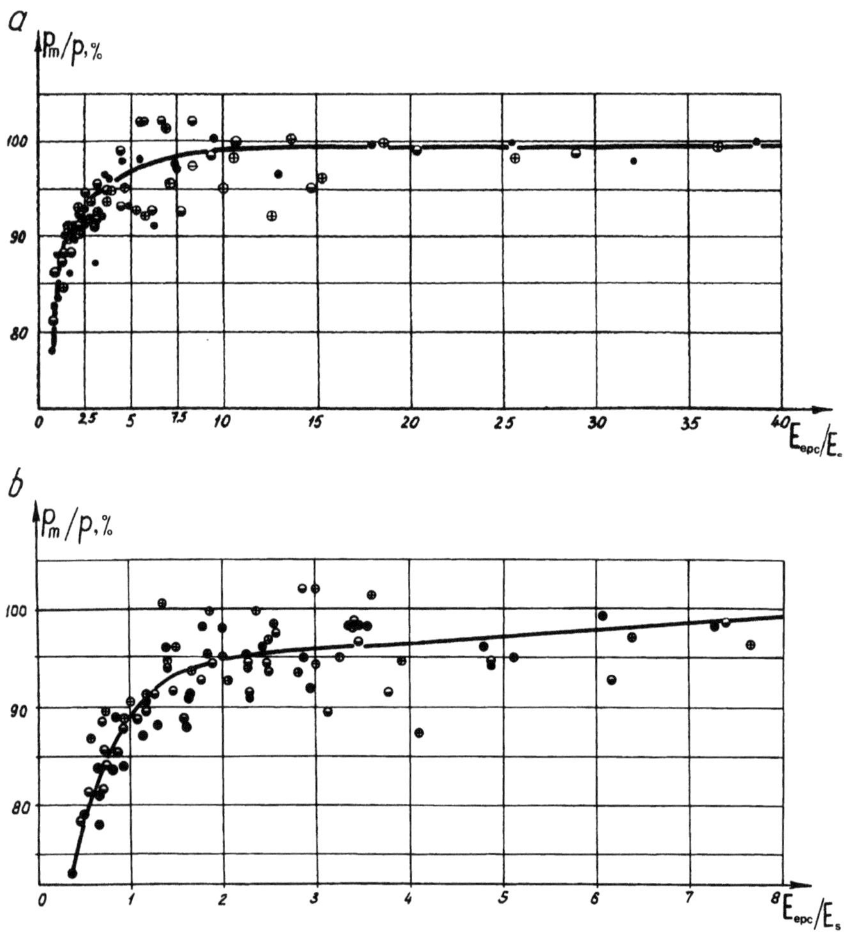

Figure 5-9. Relationship between p_m/p and E_{epc}/E_s obtained from tests with pressure cells depicted in Figure 5-1c. (a) First loading cycle; (b) second loading cycle.

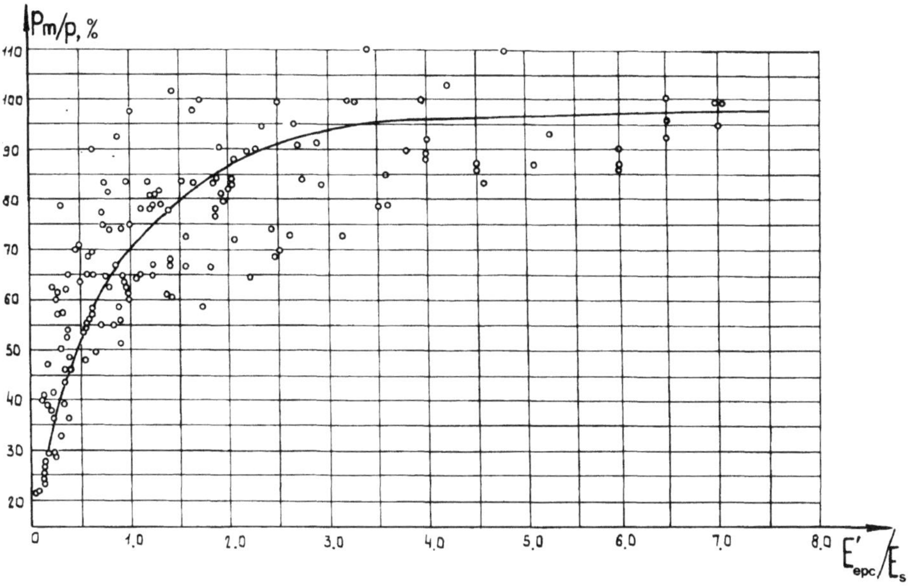

Figure 5-10. Relationship between p_m/p and E_{epc}/E_s obtained from tests with pressure cells depicted in Figure 5-1b.

from this figure, the curve that approximates the relationship between p_m/p and E_{epc}/E_s follows the same pattern as in the previously described investigation. However, in this set of experiments a decrease in error in soil pressure measurements is somewhat slower to achieve and cell readings are more scattered than those in the tests with stiffer instruments. The latter is also explained by the fact that the active soil pressure-bearing area of the cell in question is relatively small.

As noted earlier, cell deflection and its related deformability, as considered in the above experiments, are related to the maximum membrane deflection that occurs at the center of this membrane. The instrument's modulus of elasticity that results is usually referred to as the "conventional" E'_{epc}. It is of practical interest to compare the errors in soil pressure measurement made by cells in which the active membrane deflects at the center with those in which the membrane moves parallel. This was carried out on two instruments having the same values of deformability of contact membrane under similar pressure conditions. Full details of this investigation are given in Smirnov (1972). The experiments were performed with the membrane-type cells depicted in Figure 5-1c (cell Nos. 2 and 7) and in Figure 5-1b (cell Nos. 3 and 7). The deflection parameters of these cells are presented in Table 5-1.

From these experiments it was concluded that under the given conditions, such as cell diameter and stiffness of contact pressure membrane, there is no distinct difference in soil pressure measurements taken by instruments in which the membrane is displaced parallel or by those in which the deflection takes place in the middle of the pressure-bearing membrane.

These experiments also demonstrate that the error in soil pressure measurement depends on the instrument diameter. This is illustrated in Figure 5-11. It was found that soil pressures measured by cells having a membrane diameter of d = 100 mm can be almost equal to that applied (effective) at $E'_{epc}/E_s \approx 2.6$, where E'_{epc} is the conventional modulus of elasticity of a membrane-type earth pressure cell. At this ratio the error in soil pressure measurements is in the neighborhood of 5%. From this it follows that, in soils having $E_s = 10$ MPa, the aforementioned cells having $d \geq 100$ mm and a conventional modulus of elasticity $E'_{epc} = 26$ MPa (this normally conforms to cell thickness ≈ 12 mm) can be used with confidence for soil pressure measurements.

It should be noted that, in practice, soil usually has a modulus of deformation that exceeds 10 MPa. However, as discussed earlier, after repeated loading the modulus of deformation can be substantially higher, especially on unloading (Baranov 1962, 1982). During unloading, hysteresis in the cell's behavior is observed; e.g., the greater the soil modulus of deformation and the thinner the instrument's soil pressure-bearing membrane, the greater the hysteresis. The readings of pressure cells with thin membranes do not (practically) decrease at the initial stages of unloading. For example, the pressure in the calibration device

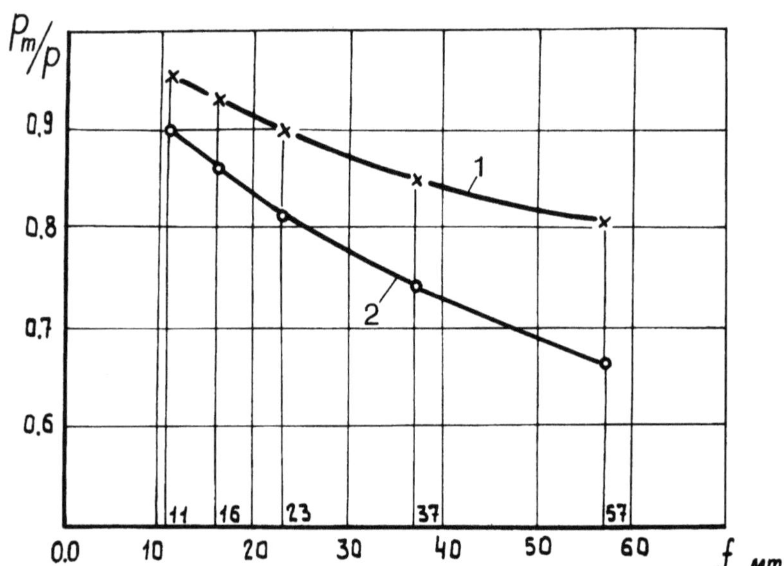

Figure 5-11. Differences in soil pressure measurements depending on diameter of soil pressure-bearing membrane. 1. Membrane d = 100 mm; 2. membrane d = 50 mm.

can be decreased down to 0.20–0.15 MPa, but the pressure cell still shows 0.3 MPa.

A strong interdependence between stresses measured by a soil pressure cell and the diameter of the soil pressure-bearing membrane of the cell was confirmed experimentally using pressure cells having identical stiffnesses (E'_{epc} = 210 MPa) but with different membrane diameters (d = 20, 30, 40, 60, and 80 mm). The results of this investigation are illustrated in Figure 5-12. From this figure it is seen that the effect of the membrane diameter shows up just slightly in soil with low moduli of deformation, e.g., E_s = 10.7 and 20.7 MPa, whereas it increases substantially in soils having high values of E_s. In the latter case the higher value of E_s typically results in an increased error in measurements of soil pressure. For

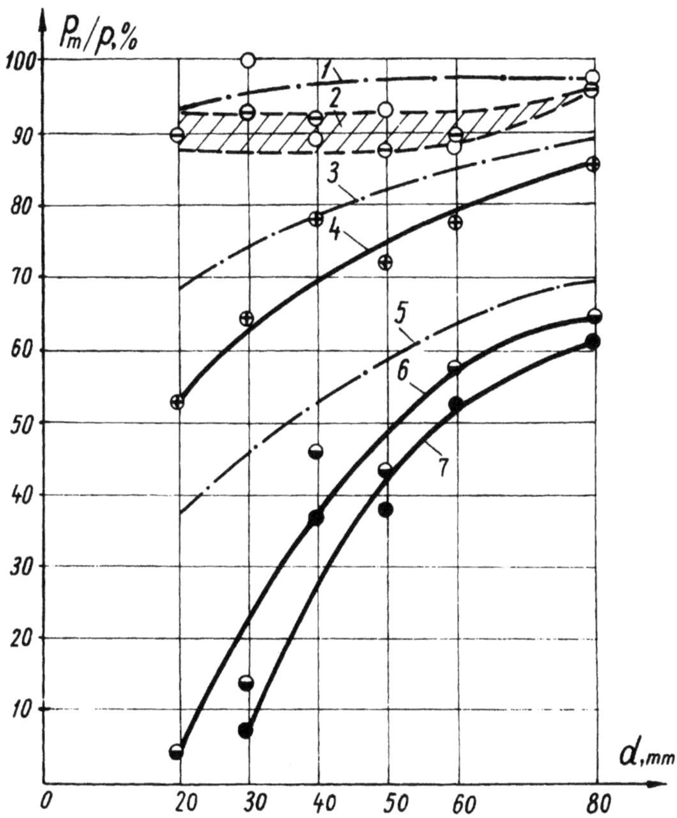

Figure 5-12. Effects of size of soil pressure-bearing membrane on measured soil contact pressures. 1, 3, and 5. Soil pressures as function of membrane diameter for E_s = 10.7 and 20.7 MPa, 110 MPa, and 400 MPa, respectively; 2. pressures measured in soils having E_s = 10.7 and 20.7 MPa; 4. measured in soils having E_s = 107, 108, and 110 MPa; 6. measured in soils having E_s = 210 and 234 MPa; 7. measured in soils having E_s = 353 and 400 MPa.

example, at E_s = 107 to 110 MPa (curve 4) and membrane diameter d = 20 mm, the error in soil pressure measurement reaches 47%, as compared to 14% for d = 80 mm. At E_s = 212 to 234 MPa (curve 6), these values are 96% and 35%, respectively. In soils having E_s = 353 to 400 MPa, the error in soil pressure measurement by cells with membrane diameter equal to 80 mm is approximately 40% (curve 7).

Errors in pressure cell performance obtained by equation (5-3) are also plotted in Figure 5-12. As is seen from this figure, the disagreement between the experimental and analytical curves is quite obvious. For soils with low moduli of deformation, this disagreement is basically qualitative. However, when the moduli of deformation of the soils increase, the shapes of the curves become similar, but the quantitative discrepancy increases substantially. For example, the difference in analytical experimental data reaches 38 to 40% in pressure cells having d = 20 and 30 mm with a soil having E_s = 400 MPa.

Comparison of the experimental data presented in Figure 5-12 with results obtained by equation (5-3) suggest that the basic idea of this formula is correct. The quantitative disagreement between theoretical and experimental results can be attributed to the following:

- Equation (5-3) considers a plane-parallel displacement of the soil pressure-bearing membrane and not a deflecting one.
- Different procedures used in the determination of E_s may result in a considerable difference in values of the modulus of deformation. This in turn may affect the values of Δ_p obtained from equation (5-3).

Based on the preceding experimental data, a modified formula (5-4) for determining the relative systematic error of soil pressure measurement, Δ_p, as a function of membrane diameter, d, and soil modulus of deformation, E_s, was proposed. This formula is recommended basically where contact pressure cells are installed with their active face flush with the footing's underside. It is also assumed that the stiffness of the material of the structure in which the pressure cell is installed (E_c) substantially exceeds the stiffness of the soil; i.e., $E_c >> E_s$.

$$\Delta_p = \frac{-1}{0.07 \dfrac{d}{f_{p=1} E_s} + a} \tag{5-4}$$

All parameters in this equation are identical to those in equation (5-3); d and $f_{p=1}$ are in centimeters and E_s is in MPa, and the factor a is equal to 2 in the first loading cycle and 3 in the second and following loading cycles.

Equation (5-4) with factor a = 3 is valid when soil lateral expansion under pressure is either excluded or substantially minimized and the soil modulus of deformation increases under pressure. In practice, such conditions may exist in the central part of the underside of a large and substantially stiff footing, or in other structures of similar nature. Equation (5-4) is recommended for preliminary estimates of the value of relative systematic error of soil contact pressure cells. Its use, however, does not exclude calibration of pressure cells in calibration devices prior to their installation in a structure or in a model.

5.7 Soil Stress State Investigation at Soil Pressure Cell Contact Surface

As pointed out by Selig (1964), the variation in soil/cell stiffness ratio may result in substantial errors in readings by these cells. Until relatively recently the variation in the mass of soil located in close proximity to the contact pressure cells installed in a structure, as well as the stress distribution at the cell/soil interface and in the mass of soil, was determined analytically on the basis of miscellaneous theories.

A series of experiments were conducted by this author and co-workers in order to investigate the effects of embedded contact pressure cells on the soil's stress state located in close proximity to the cell, as well as stresses at the soil/cell interface. The investigation was carried out in calibration devices such as those described in Chapter 3. The soil contact pressure and soil pressure distribution in the soil mass at some distances from the active surfaces of the soil pressure cells (e.g., 20 and 100 mm) were measured by a variety of instruments, e.g., the cells depicted in Figure 5-1c, small-diameter ($d = 29$ mm), flat, single-membrane sensors, a multicell, large-diameter instrument (Figure 5-2), and miniature flat, double-membrane pressure cells.

Loose medium sand was used in this investigation. The stress distribution in the soil mass was measured by the least and most stiff cells of the types depicted in Figure 5-1c (cell Nos. 1 and 7). The number and arrangement of miniature pressure cells placed within the masses of soil was such to ensure that average reading of soil pressure at equivalent locations was obtained from four instruments. (This does not apply to the central point.) Each set of experiments was repeated 3 to 4 times in order to ensure consistency of obtained results. The soil surface was loaded by a uniform load of $q = 0.2$ and 0.4 MPa. The characteristics of the soil stress diagram for these loads are very similar; however, scatter results of the soil stress measurements at $q = 0.2$ MPa were greater than when the soil was loaded by $q = 0.4$ MPa.

In Figure 5-13 the results of the measurements of soil stresses are given in percents assuming the nominal stress $q = 0.4$ MPa as 100% (reference). Figure 5-13a illustrates the results of measurements of stresses in a mass of soil taken by cell No. 1. From this figure it is seen that at close proximity to the instrument there is some (≈8%) stress concentration in the area of the cell's perimeter rim. At the same time, in this layer of soil, there is a tendency for stress reduction to approximately 10% in the areas that deflect away under the applied pressure.

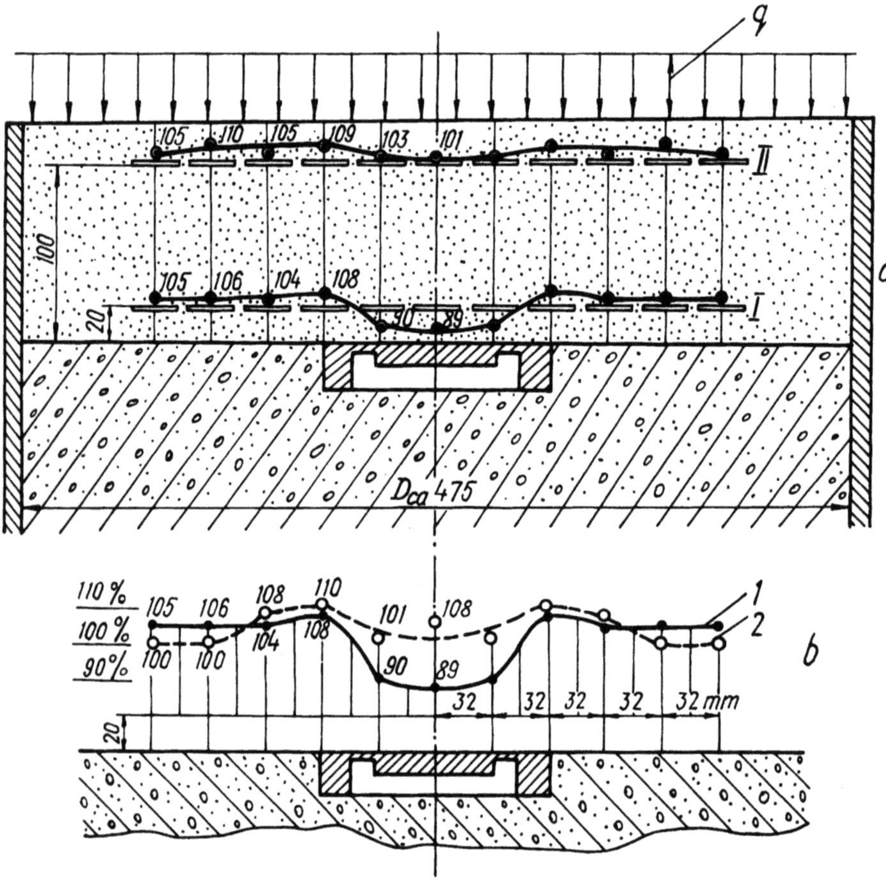

Figure 5-13. Character of stress distribution over contact surfaces of pressure cell depicted in Figure 5-1c. (a) Cell No. 1; (I) stress distribution in the immediate vicinity of active surface of centrally located pressure cell; (II) same at distance of 100 mm from active surface; (b) comparison of identical soil stress distribution diagrams obtained under similar conditions by using cells of different stiffnesses. 1. Measured by cell No. 1; 2. measured by cell No. 7. *Note:* Readings of pressure cells are presented in percents as related to the nominal pressure $q = 0.4$ MPa.

At soil levels located 100 mm above the cell's active surface the difference between the increase and decrease in soil mass stresses is less pronounced and constitutes approximately 8% (total).

A comparison of the results in stress measurements in soil masses at 20 mm above the cell's active surface by cell No. 1, and the stiffer cell No. 7 (parameters of these cells are given in Table 5-1), are presented in Figure 5-13b. From this figure it is seen that the reduction in stresses due to deflection of the cell's active surface is less pronounced for cell No. 7 than it is for the more flexible cell No. 1. It should be noted that no substantial deviation from nominal stress (that is q = 0.4 MPa) was measured by cells at the 100-mm level above cell No. 7. In conclusion, it should also be noted that the contact stresses measured by cells No. 1 and 7 are in good agreement with those depicted in Figure 5-13.

A detailed investigation of a similar nature was performed using the large-diameter multicell instrument illustrated in Figure 5-2. A cross section of the experimental setup used in this investigation is shown in Figure 5-14. As is seen in this figure, the instrument's pressure cells are part of the large-diameter membrane and are installed flush with the concrete base. Miniature cells were used for measurement of stresses in the soil mass above the large-diameter instrument, arranged in layers—one layer at a time in each experiment. The instrument was calibrated by air pressure (p) equal to 0.2 and 0.4 MPa. This was considered as equal to 100% (reference) in the subsequent investigation. Medium sand (loose D_r = 30% and dense D_r = 67% was used in these experiments.

The purpose of this investigation was to determine the character of stress distribution in the soil mass, and in particular, in the soil zone above the large-diameter instrument's soil pressure-bearing surface. Some characteristic results of the experiments are illustrated in Figure 5-15. Distribution of the soil stresses

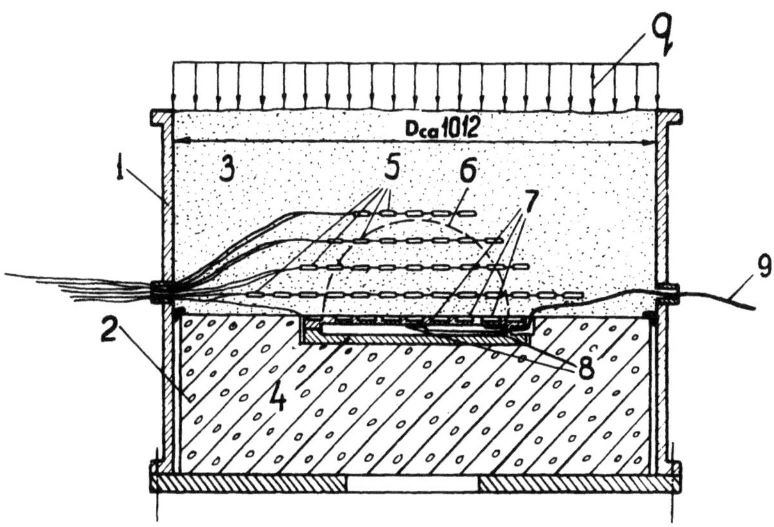

Figure 5-14. Experimental setup for investigation of effects of deformability of the active surface of soil pressure cell on stress distribution within mass of soil. 1. Test chamber; 2. concrete base; 3. soil; 4. large-diameter multicell instrument; 5. small cell; 6. soil zone affected the most by deflection of instrument's membrane; 7. soil pressure cell included in large-diameter instrument; 8. resistance strain gages mounted on back of membrane; 9. cables. *Note:* Only one layer of miniature cells (5) was used in each set of experiments.

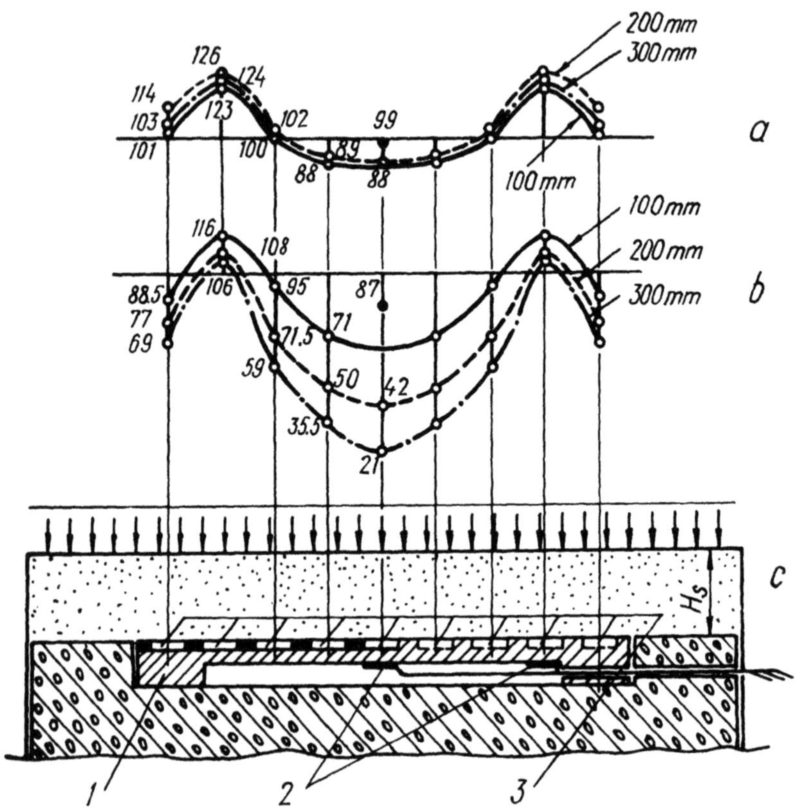

Figure 5-15. Contact pressure distribution in mass of soil above large-diameter soil pressure cell in percents as related to "nominal" stress q = 0.4 MPa. (a) In loose sand; (b) in dense sand; (c) typical cross section of experimental setup. 1. Body of large-diameter cell; 2. resistance strain gages; 3. small soil pressure measurement cells. *Note:* Measurements taken at 100, 200, and 300 mm above instrument's active face.

were measured at different heights, e.g., 100, 200, and 300 mm above the large-diameter soil pressure cell. The values of the stresses measured, presented in Figure 5-15, are given in percents related to the "nominal" stress readings obtained by miniature cells when these cells were loaded by the compressed air load q = 0.4 MPa.

As is seen from Figure 5-15, the contact stress distribution within the soil mass is very similar to that described in Chapter 1. It is characterized by an increase in value over the stiff (nonyielding) parts of the soil contact pressure measuring instrument installed within a stiff bottom (insert) and decreases where the active soil pressure-bearing membrane deflects; at all layers where the stresses were measured qualitatively the character of stress distribution is similar. Interestingly, in the case of loose sand, the difference in soil pressures at the centerline of the instrument was only 1% (Figure 5-15a).

Simultaneously with the contact stresses measured in soil mass, readings were taken from the strain gages attached to the backside of the soil pressure-bearing membrane of the instrument installed at the bottom of the chamber. The contact stresses measured by this instrument correlate very well with those measured in the soil mass, as previously discussed.

It should be noted that the stiffness of the soil pressure-bearing membrane is fairly low; as noted earlier, the elasticity modulus of the instruments, E_{epc}, is equal to 72.5 MPa. This qualifies this instrument as low to medium stiff. In medium sands exposed to stresses p equal to 0.2 and 0.4 MPa, the E_{epc}/E_s ratio varied from 0.3 to 0.7. Here E_s is the soil modulus of deformation.

Figure 5-16 illustrates the results of similar tests that were carried out with loose medium sand having a relative density of $D_r < 30\%$. In this series of experiments the depth of soil in the calibration device, H_s, was equal to 400 mm, and the soil pressure was measured at distances (H_1) from the active face of the large-diameter soil pressure measuring instrument that varied from 50 mm to 350 mm. As in other tests, the large-diameter cell was installed with its soil pressure-bearing membrane flush with the surface of the concrete insert. For greater confidence in the results, all tests were repeated twice. Characteristic for all measured stresses at all levels in the soil mass was a decrease in soil pressures above the deformable part (membrane) of the large-diameter pressure measuring device. This phenomenon was more pronounced within the soil mass located closer to the instrument's active face (membrane). Small cells installed on the outside surface of the soil pressure-bearing membrane and resistance strain gages glued to the inner surface of this membrane also registered a decrease in soil pressure (as compared to the readings during calibration by compressed air) at the interface between the soil and instrument (Figure 5-16c).

At the membrane's stiff support, increases in soil pressure were registered. This can be partly attributed to the membrane performance and its behavior as a thin plate fixed at all edges. From Figure 5-16a it is seen that in a loose sand at $H_1 = 300$ mm and up, deflection of the membrane has no practical effects on soil pressure measurements within the mass of soil. However, in compacted sand ($D_r > 67\%$) this phenomenon is less pronounced. One explanation for this is that in loose sands the zone of deformed soil over the deflecting membrane is substantially smaller than in a dense sand. This is also typical for instruments with an active surface that is displaced parallel to its original surface while exposed to the pressure. In this case, however, the depth of the soil-deformed zone is somewhat smaller than in the former case, i.e., the membrane fixed at the instrument's perimeter rim.

To complete the discussion on systematic errors in soil measurements by pressure cells, the last case is that of soil pressure measurements made in moist

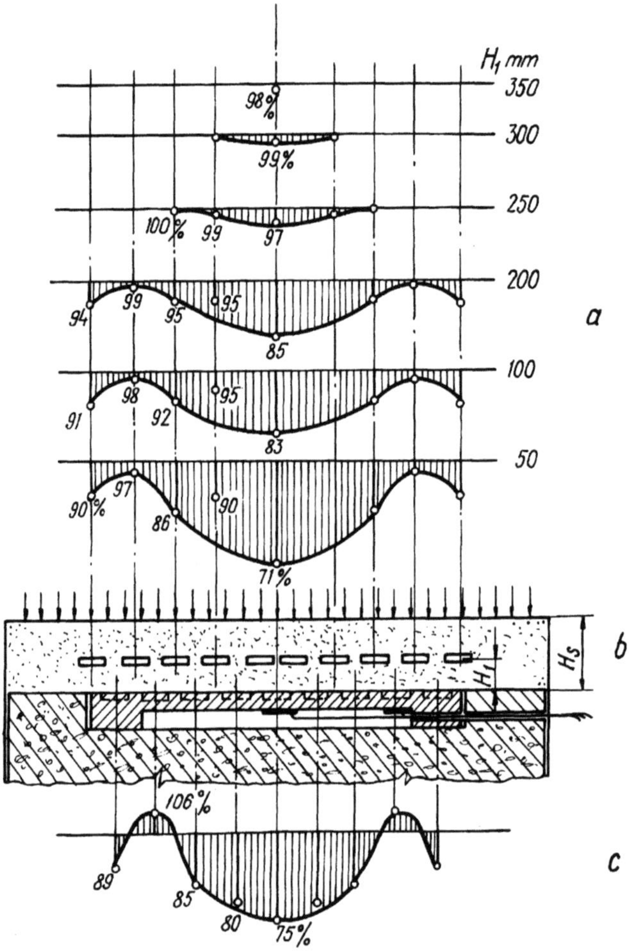

Figure 5-16. Stress distribution in mass of loose sand and at soil pressure cell interface. (a) In soil mass as a function of H_1; (b) typical cross section of experimental setup; (c) distribution of contact pressure at interface of large-diameter pressure cell.

loam having a disturbed matrix, using the pressure cells illustrated in Figures 5-1a and 5-1c. Soil particle distribution data and other relevant characteristics are given in Tables 5-5 and 5-6.

The results of these tests are presented graphically in the form p_m/p (%) $= f(E_{epc}/E_s)$ in Figure 5-17. From this figure it is possible to conclude that the stress distribution in a cohesive soil of the type used in these experiments is very close to linear ($p_m \approx p$). More information on this subject is found in works by Trollope and Lee (1961) and Baranov et al. (1971b).

Table 5-5. Grain Size Distribution and Other Relevant Characteristics of the Loam Used in Soil Pressure Experiments

Soil Particles (mm), in Percent of Weight						Density of Dry Soil ρ_d, kN/m^3		Void Ratio (e)		Water Content
1–0.5	0.5–0.25	0.25–0.05	0.05–0.01	0.01–0.005	Less than 0.005	Loose	Dense	Max	Min	w, %
1.2	2.2	17.9	64.5	5.9	8.3	12	1 7.5	1.20	0.59	7.1

Table 5-6. Moduli of Deformation of the Loam Obtained Experimentally

		Moduli of Deformation E_s at Pressure p, MPa			
Soil Condition	Loading Cycle	0.1	0.2	0.3	0.4
Loose	First	1.0	2.0	3.4	4.5
	Second	62.0	82.5	115.0	91.0
Dense	First	26.5	28.0	29.5	34.0
	Second	78.0	89.0	149.0	152.0

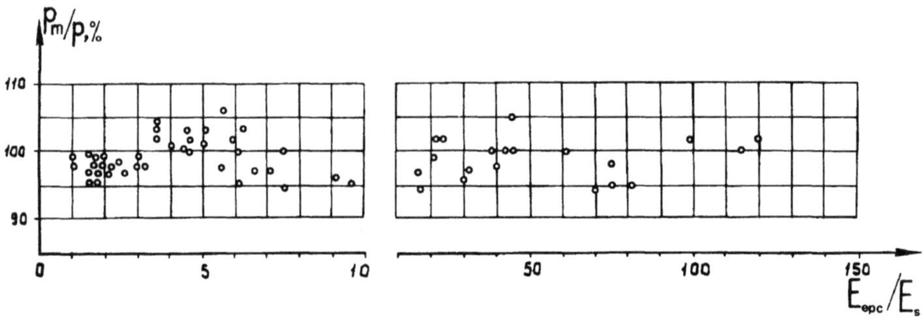

Figure 5-17. Relationship between p_m/p (%) and E_{epc}/E_s in cohesive soil with disturbed matrix.

5.8 Results of Investigation of Systematic Relative Error in Stress Measurements in Soil Performed with Nonstiff Instruments

The results of theoretical and experimental investigations carried out by different workers, as well as test results presented in this and earlier chapters of this work, have demonstrated a strong interdependence and complexity of interaction between the soil and nonstiff pressure measuring cells. The following are basic factors that affect the soil/nonstiff instrument interaction, provided that the instrument is installed absolutely flush with the interface:

- type of soil
- diameter of contact (active) membrane diameter, *d*
- character and value of displacement of soil pressure-bearing surface
- soil modulus of deformation
- direction of applied pressure
- space between adjacent instruments.

The experimental results have shown that there is a strong interdependence between the nonstiff cell diameter, *d*, and the error in the measurements of soil pressure. For example, under similar soil and loading conditions the difference between calculated errors in soil measurements by instruments having diameters 50 and 100 mm ($\Delta_{p(d=50)}$ and $\Delta_{p(d=100)}$) would be approximately 4.7 to 13.4%. In instruments with smaller diameters, e.g., $d = 20$ and 30 mm, having similar moduli of elasticity as in the above example, the calculated error in soil pressure measurements, Δ_p, in soils having $E_s = 353$ and 400 MPa, respectively, is 87 and 97%. This strongly suggests that instruments with relatively flexible soil pressure-bearing membranes are, in general, unsuitable for measurement of soil contact pressure in dense cohesionless soils. On the basis of the earlier described experiments, it is possible to conclude that small-diameter flexible membrane pressure cells can be used for pressure measurements in soils having $E_s \leq 10$ to 20 MPa.

One, however, must be aware that the error in these measurements can be substantial and can exceed 10%.

As noted earlier, this can be explained by the formation of a dense nucleus in a zone of soil just above the instrument's membrane. The height of this zone can reach an approximate value of $3d$. The calibration curves of cells having flexible membranes is usually nonlinear. This suggests that the character of soil compressibility above a flexible active surface is not constant and is usually affected by the membrane's deflection, f.

Stiff membranes produce much better calibration results. Therefore, to obtain credible results in soil pressure measurements, the membrane's deflection (stiffness) should be limited to a certain value. Trollope and Lee (1961) recommend that the membrane deflection, f, at a pressure of 0.1 MPa should not exceed 4 micrometers. This recommendation has been confirmed in part by an investigation conducted by this writer and co-workers. In this investigation it was found that the recommendation by Trollope and Lee is valid if pressure cells are used to measure contact pressures in loose ("not stiff") soils; much lower values of f are required in dense ("stiff") soils. This fact is also important where it is necessary to take readings in unloading and in subsequent loading cycles.

This suggests that for reliable results the limitations on deflection of the soil pressure-bearing surface of the pressure cells must be considered. Results of the investigations described in this and previous chapters suggest that to achieve a minimum systematic error in measuring the soil contact pressures, the ratio E_{epc}/E_s should not be less than 7 for a pressure cell having $d = 50$ mm and not less than 2.6 for a cell with $d = 100$ mm. Here E_{epc} and E_s are, respectively, the modulus of deformation of the membrane-type instrument and the soil. Under the above cell/soil conditions, and also where the instrument's thickness does not exceed 12 mm, the systematic error in the measurement of soil pressures does not exceed 5%.

Acceptable (recommended) values of membrane deflection at its center, when the above conditions are considered, are presented in Table 5-7. It can be seen that in dense (stiff) soils the permissible deflection of the cell's active surface is very small. Especially sensitive transducing systems are required to measure such small deflections.

The systematic error in soil pressure measurements by instruments having small-diameter pressure-bearing membranes grows substantially, and the scatter in readings by these instruments increases dramatically. In this writer's opinion, instruments having $d = 100$ to 120 mm may represent the optimal solution for membrane-type soil pressure measuring instruments. Also, instruments of this type should not be used in dense cohesionless soils having $E_s > 40$ MPa.

In cohesive soils, because of their compressibility characteristics, the limitations applied to soil pressure measuring cells are less stringent than those in cohesionless soils. The investigations have shown that in most practical cases cohesionless soils can be treated as very viscous liquids.

In all types of soils the use of instruments having the pressure-bearing surface displaced parallel under soil pressure is preferable. The more complex structure

Table 5-7. Recommended Deflections (Maximum) at Centers of Pressure-Bearing Membrane of Membrane-Type Soil Contact Pressure Cells as Required to Meet Condition $\Delta_p \leq 5\%$

Modulus of Soil Deformation, E_s	Required Nominal Modulus of Elasticity, E_{epc}, of Membrane-Type Pressure Cell, MPa		Deflection, f, at Center of Membrane at Pressure Increment 0.1 MPa, μm	
	d = 50 mm	d = 100 mm	d = 50 mm	d = 100 mm
20.7	145	53.8	8.3	22
40	280	104	4.3	11.5
110	770	286	1.6	4.2
147	1029	382.2	1.2	3.1
234	1638	608.4	0.7	1.9
400	2800	1040	0.4	1.2

(and therefore higher cost of manufacturing) of these instruments is compensated by smaller values of errors in the measurement of soil pressures. This conclusion is also supported by information found in published and unpublished reports produced by many investigators, e.g., Uginchus and Bombchinsky (1954), Kallstenius and Bergau (1956), and others, as well as by results of the earlier discussed experiments.

And again, one should realize that substantial parallel displacement of the soil pressure-bearing membrane of the contact pressure cell, relative to the surface of the structure in which it is installed, can result in substantial scatter in instrument readings even under homogeneous soil conditions. This is a manifestation of the "edge effect" between the membrane and the surface of the structure. Obviously, in the case of a "flexible" contact soil pressure cell, the combined form of cell's active surface deformability, as is shown in Figure 5-1c, has the advantage in comparison with an instrument having a "pure" parallel displaced soil pressure-bearing membrane.

An instrument based on these principles is shown in Figure 5-18. The operation

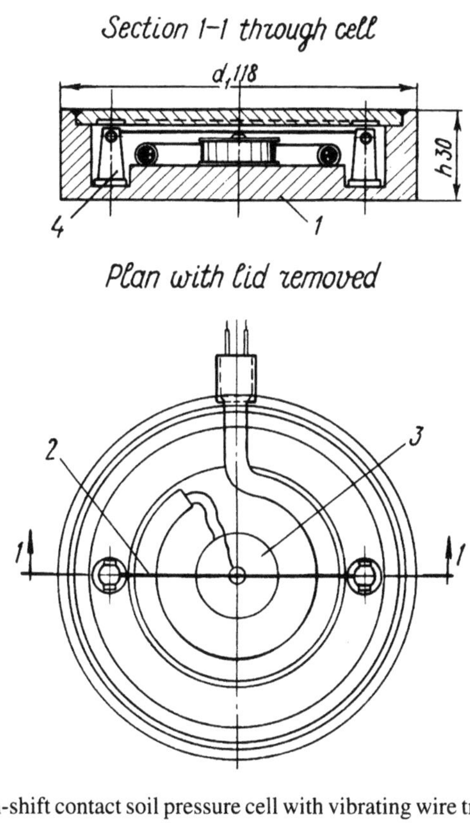

Figure 5-18. Non-shift contact soil pressure cell with vibrating wire transducer (Uginchus and Bombchinsky 1954). 1. Contact surface; 2. sound wire; 3. electromagnet; 4. bracket.

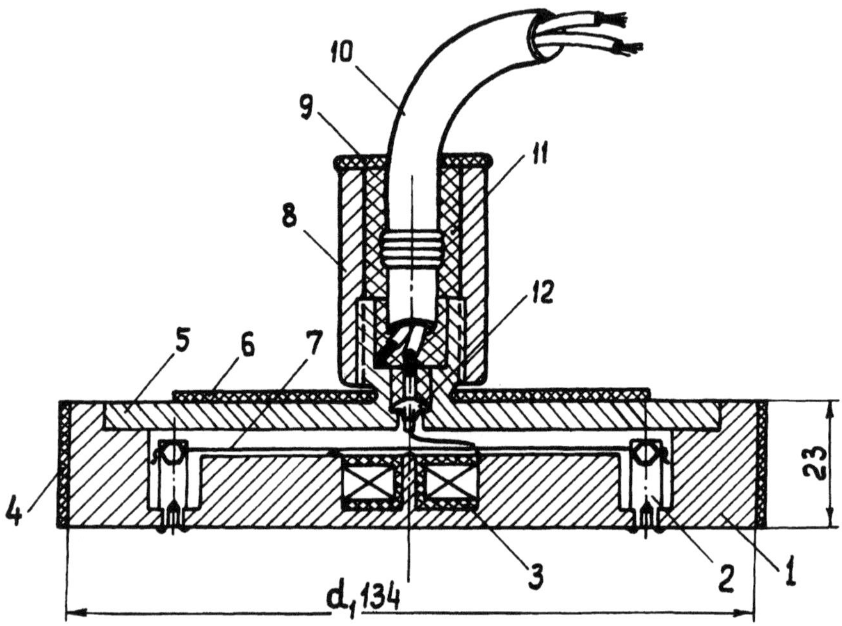

Figure 5-19. Improved version of pressure cell shown in Figure 5-18. 1. Body; 2. bracket; 3. electromagnet; 4, 6, and 9. soft material cover; 5. lid; 7. vibrating wire; 8. pipe connection; 10. double-core cable; 11. epoxy-based cement; 12. insulator.

of this instrument is based on changes in a wire vibration frequency due to deflection of the soil pressure-bearing membrane. Membrane-induced tension to the wire is transmitted via brackets that are welded to the instrument's contact membrane.

A modified version of this type of cell is shown in Figure 5-19. Some parameters of this instrument are given in Table 5-8. Investigation of this instrument's

Table 5-8. Parameters of Vibrating Wire Type Soil Pressure Cell Illustrated in Figure 5-19

Maximum Pressures Measured by Pressure Cell (MPa)	Calculated Modulus of Elasticity of Pressure Cell Body E_{epc}, MPa	Pressure Cell Sensitivity (per 1 Hz)	
		Percent of Measuring Range	MPa
0.05	100	0.8	0.0004
0.1	165	0.6	0.0006
0.2	235	0.5	0.0010
0.3	475	0.5	0.0015
0.5	770	0.5	0.0025
1.0	910	0.5	0.0050
2.0	1150	0.5	0.0100

performance indicates that, having E_{epc} = 100 MPa, it can be used to measure the contact pressures in loose and dense medium- and coarse-grain sands having $E_s \leq 55$ MPa during the first loading cycle. The expected error in soil pressure readings, Δ_p, can be in the neighborhood of about 6 to 8%. As pointed out earlier, the repeated loading of a granular soil and of dense sands in particular, under confined conditions results in substantial increases in E_s. This, in turn, results in an increased error in cell reading; e.g., in the second loading cycle of a dense medium sand $\Delta_p \geq 18\%$. In practice, such conditions exist in the conventional calibration chamber or at the central portion of a large foundation slab.

6

Full-Scale and Field Experiments: Monitoring of In Situ Soil/Structure Interaction

Full-scale and field experiments are usually an expensive and time-consuming effort. To be successful they require a thorough and comprehensive preparation that includes a definition and objectives of the investigation and proceeds through a series of steps with regard to instrument specifications, arrangement, readings taken, and processing of the obtained data. In practice these experiments cannot be repeated, so all efforts must be considered to reveal and reinforce any weak links that can contribute to the failure of the investigation.

As pointed out by Franklin (1977), the field monitoring program is a chain with many potentially weak links. Breakdowns occur with greater frequency in this area than in most other tests in geotechnical engineering. Hence, it is of paramount importance that investigation or monitoring program planning and conducting is carried out by personnel with specialist expertise in the application of geotechnical instrumentation in general and in soil pressure-measuring instruments in particular. For detailed discussions on a systematic approach to planning field monitoring programs using geotechnical instruments, the reader is referred to work by Dunnicliff (1988).

6.1 Arrangement (Layout) of Pressure Cells and Installation Procedures

Preparation for any kind of investigation, and for full-scale or field experiments in particular, includes the careful formulation of the objectives of these experiments (monitoring program) and a clear understanding of the means by which these objectives can be achieved. The latter usually includes selection of the type and number of relevant instruments and their placement at locations critical to the performance of the structure (model). The selection of the number of instruments and their location within the investigated or monitored structure usually depends on a predicted interaction between the structure and the foundation soil. In modern practice finite element analysis is usually used for this purpose. As pointed out by Dunnicliff (1988), a practical approach to selecting instrument locations and their layout entails the following three steps:

- Zones of particular concern are identified.
- If there is no zone of particular concern and instruments are to be located elsewhere, then a typical cross section, where predicted soil/structure interaction is considered as representative, is selected.
- Because the selected zones or cross sections considered as primary locations may be incorrect, zones of secondary importance are selected. This will provide the investigator(s) with a confidence in the final outcome of his/her investigation.

When designing the instrument layout pattern the survivability of the instruments in their intended environment should be considered. In the case of important investigations or where instruments are used for monitoring the performance of important structures, installation of additional instruments may be warranted to replace instruments that may become inoperative. In certain cases loss of the function of these instruments can be substantial.

According to Abramson and Green (1985), the average survivability rate for load cells can be 75%, whereas for strain gages it is 60%. The latter number is also reported by Tsinker (1997). In general, in full-scale and field investigations it is unwise to rely on a single instrument as an indicator of model or structure performance; the number of instruments installed and their layout should provide for a cross-check of instrument readings. The number of relevant examples are discussed elsewhere in this book. Finally, if warranted, in addition to prime objectives of the investigation, the experiments of secondary importance, e.g., testing of different soil pressure-measuring systems, measurement of not only contact soil pressures but also distribution of stresses in the soil mass below the soil/structure interface or in the backfill material, etc., should also be considered.

What follows is a discussion on several characteristic full-scale experiments carried out by this writer and co-workers. A relevant example on the investigation of two instrumented rectangular foundation blocks was discussed in Chapter 2. As was pointed out, the purpose of conducting these tests on two large-sized rectangular concrete blocks was to compare two techniques of measurement of contact stresses. In one of these techniques the contact pressure cells were installed flush with the concrete face, and in the other technique stiff, flat pressure cells were installed in the mass of foundation soil in close proximity to the underside of the test block. The layout of the installed contact pressure cells is given in Figures 2-1 through 2-3. The number of instruments used in these experiments was designed to generate reliable data during the investigation.

In the stiff concrete block having a width-to-length ratio approximately equal to 1:4 (Figures 2-1a and 2-2), the rows of contact pressure cells are located in the middle portion of the block. This is done to avoid the potential "end effect" on the results of the stress measurements. Cells that were installed in the mass of foundation soil were also arranged under the middle part of the test block for comparison of the obtained results. In an effort to find the practical distance limit between the instruments installed within the mass of soil and the footing's underside, these instruments were installed at three different parallel levels in the mass of soil (Figure 2-1).

Furthermore, an additional six cells were installed at the block ends and four additional cells were placed at the block centerline. This was done to investigate the stress distribution at the block ends and for additional control of test results. Secondary purposes of these investigations included the testing of the operational reliability of two different transducing systems as related to the given test conditions, scatter in cell readings depending on cell diameter, effect of cyclic loading on soil/structure interaction, and settlement of the test block under different load conditions. For details see Chapter 2. This particular example demonstrates how several problems can be investigated in one specific set of tests. Similar examples carried out on a circular concrete slab are also found in Chapter 2.

The aforementioned case is not a standard example of an investigation of soil/structure interaction. In practical cases the number of instruments installed for monitoring the interaction between the foundation soil and the structure is usually limited. As pointed out earlier, the pressure cells are to be installed at specific locations where the expected soil pressure can change its value substantially. The number of instruments installed at these and other locations depends on the desired accuracy of the subsequently drawn soil pressure diagram. This is demonstrated in Figure 6-1. In this case the pressure differentials ($p_2 - p_1$) are at a maximum at locations *n* of the theoretically determined soil pressure diagram, and at a minimum ($p_4 - p_3$) at location *c*. Hence, to obtain a more accurate soil pressure diagram, more instruments are required at locations *n* than at location *c*.

It should be noted that the shape of the soil reaction diagram depends on many factors: footing stiffness, value and character of load application, type of foundation soil, and others. Hence, theoretical determination of the soil pressure diagram must be given proper attention during execution of the investigation program. The example of a contact pressure cell arrangement in a concrete test slab that is close to square (width-to-length ratio $\approx$ 1.4) is shown in Figures 2-1b and 2-3.

Tarikuliev and Murzenko (1969) successfully used a diagonal arrangement of contact soil pressure cells (Figure 6-2) to study a stiff square footing/soil interaction. The instrument layout was developed on the basis of a carefully determined theoretical diagram of soil reaction. This cell arrangement resulted in readings, for each characteristic point (except for the central one), being doubled four times.

Usually construction contractors ae responsible for providing support services during implementation of the instrumentation program; however, they do not

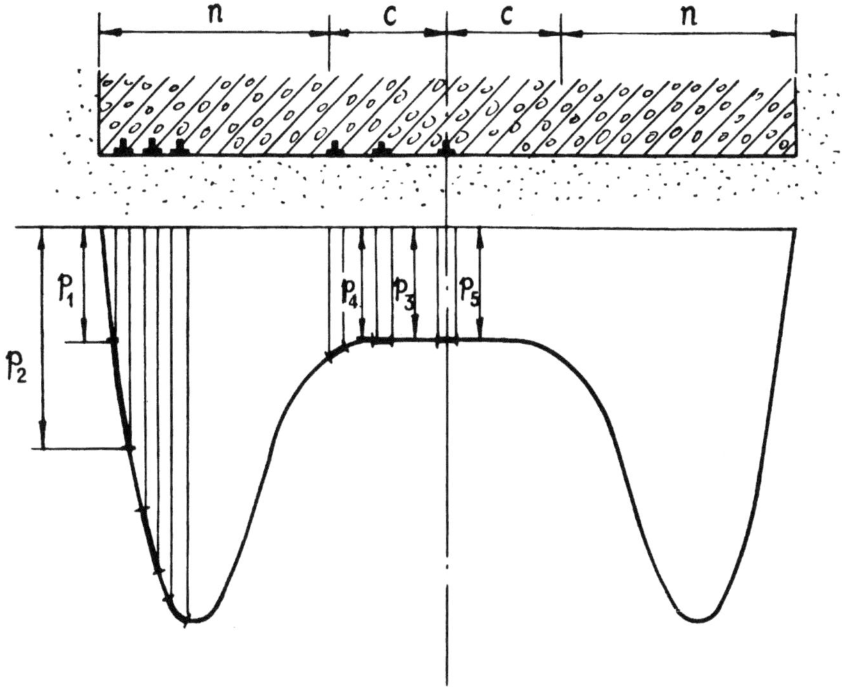

Figure 6-1. Location of soil pressure cells in footing. Typical cross section and theoretical soil reaction diagram.

have the economic or professional incentive to contribute toward the planning, investigation, or monitoring program, and clearly have no responsibility for the instrumentation program. Contractors often regard the instrumentation program as an interference with normal construction work adversely affecting construction schedules. Therefore, to prevent damage to instruments and the network of cables that may be present during instrument installation and data collection phases, the system should preferably be under the tight control of personnel authorized by the owner, instrumentation specialists, and/or the design consultant. Experience, however, indicates that, in spite all the precautions, damages to instruments and their connecting cables do occur and, as noted earlier, a great percentage of instruments is lost during construction activities.

As the overriding feature of the expensive field investigation (or monitoring) program is reliability, the program designer should be sure that failure of one or several instruments will not result in complete failure of the whole program. To ensure that this does not occur, in a full-scale (field) investigation or monitoring program, the instruments should be installed at several locations having identical soil/structure interaction conditions. In general, instruments should be installed

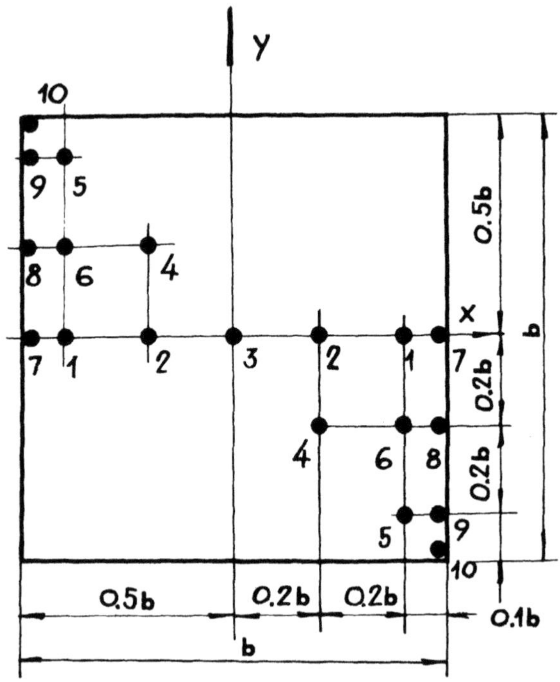

Figure 6-2. Layout of contact soil pressure cells in square footing (after Tarikuliev and Murzenko 1969).

in at least three locations (sections) having similar performance parameters. Additional locations should also be considered to duplicate instrument readings taken at the primary locations.

Typically, the number of instruments at one particular location is determined by the complexity of the foundation soils, the type of structure, and the properties of the backfill soils. For example, in a gravity-type soil retaining wall, depending on the wall height, the contact pressure cells in the upper part of the wall can be installed at 3- to 5-m spacings; in the lower part the spacings should be reduced to 1.5 to 2 m.

Naturally, each case is site specific and only general recommendations can be offered on the number and layout of instruments to be used in a full-scale and/or field investigation (monitoring) program. Drawing from his personal experience and numerous publications concerning the subject matter, the writer offers the following basic recommendations:

- If the experiment can be repeated, as it was in the aforementioned tests using instrumented large concrete blocks, then the test program may be divided into two phases. During the first phase instruments can be installed uniformly, and during the second phase instruments should be rearranged as required to obtain more accurate soil reaction diagrams, e.g., close to each other where stress concentrations take place and at greater distances where the stress diagram is relatively flat (the example is depicted in Figure 6-1).

- If the experiment cannot be repeated, that is, in the case of a field investigation, then it is recommended that instrument installation be done based on a theoretically predicted soil reaction diagram. This is best determined by the finite element method. The calculations must be based on the best available information on geotechnical properties for the founding soils and accurate specification of boundary conditions.
- In areas of stress concentration the instruments should be placed as close as possible, however, not closer than $2d$ (center-to-center) of the instruments. This is required to avoid interaction between adjacent cells.
- Reliability of obtained data from a field investigation is a must. Hence, the instruments should be installed at several identical locations, say at least two, where soil pressure measurements will be taken in at least three typical cross sections.
- Special care must be exercised during installation of contact pressure cells in structures to be monitored or investigated. In the case of a cell designed for measurement of the contact soil pressures at the underside of a footing, it must be carefully installed on a thoroughly leveled foundation soil and slightly rubbed in. The installer must understand that eventually the cell's contact surface must be flush with the underside of the footing. Subsequently a layer of concrete having very fine aggregates and layered no thinner than 100 mm should carefully be placed on top of the cell and slightly densified by light blowings made either by hand or wooden bar. The contact cell installation is best achieved by installing cells in prefabricated small blocks that are subsequently incorporated in the foundation structure. These blocks are shown in Figure 8-6.
- In the case of vertical or inclined walls the instruments can be fastened to the inside surface of the wooden formwork by means of special clamps. An example is shown in Figure 6-3. If reinforced-concrete formwork is considered, then the pressure cells can be installed flush with the external surface of the formwork panels. It should be mentioned that the method of instrument installation should be carefully reviewed, as it is site specific.

A practical example of a contact pressure cell distribution (arrangement) for monitoring the interaction between a solid type construction of navigation lock with foundation and backfill soils is shown in Figure 6-4. In this case the cells' arrangement is based on the expected stress distribution at the soil/structure

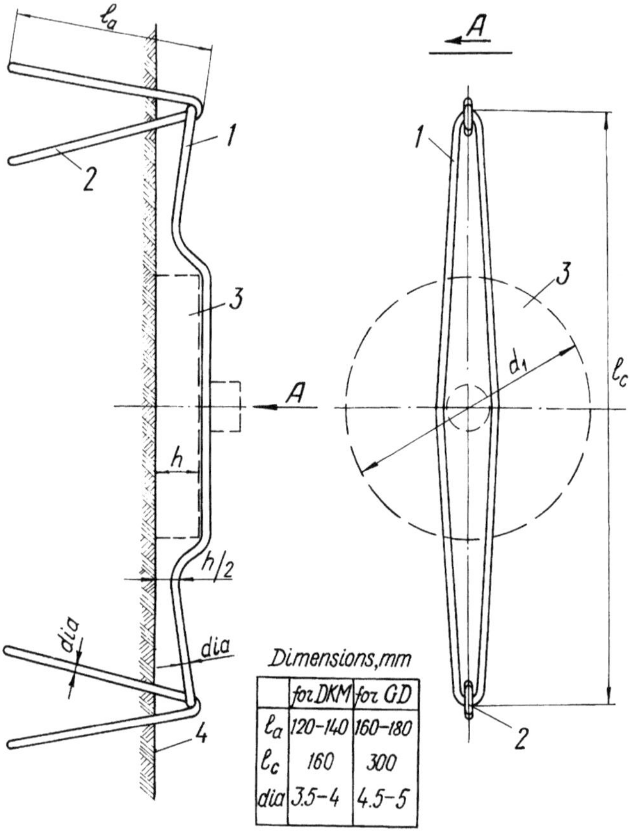

	for DKM	for GD
ℓa	120–140	160–180
ℓc	160	300
dia	3.5–4	4.5–5

Figure 6-3. Spring clamp used to hold instrument against stiff cohesive soil. 1. Clamp; 2. holding bar; 3. pressure cell; 4. surface of stiff cohesive soil.

interface. This was obtained after consideration of the elastic properties of the foundation soils and miscellaneous design load combinations. Detailed guidelines on the installation of instruments used in full-scale and field investigations are given in Uginchus and Bombchinsky (1954), Baranov (1982), and Dunnicliff (1988).

6.2 Effects of Misalignment of Cell Contact Surface with Surface of the Structure

As discussed earlier, the best results in contact soil pressure cell performance are obtained when the cell's active face is absolutely flush with the interface. In some instances, however, for various reasons the cell's pressure-bearing membrane is located either below or above the interface. This writer, with co-workers,

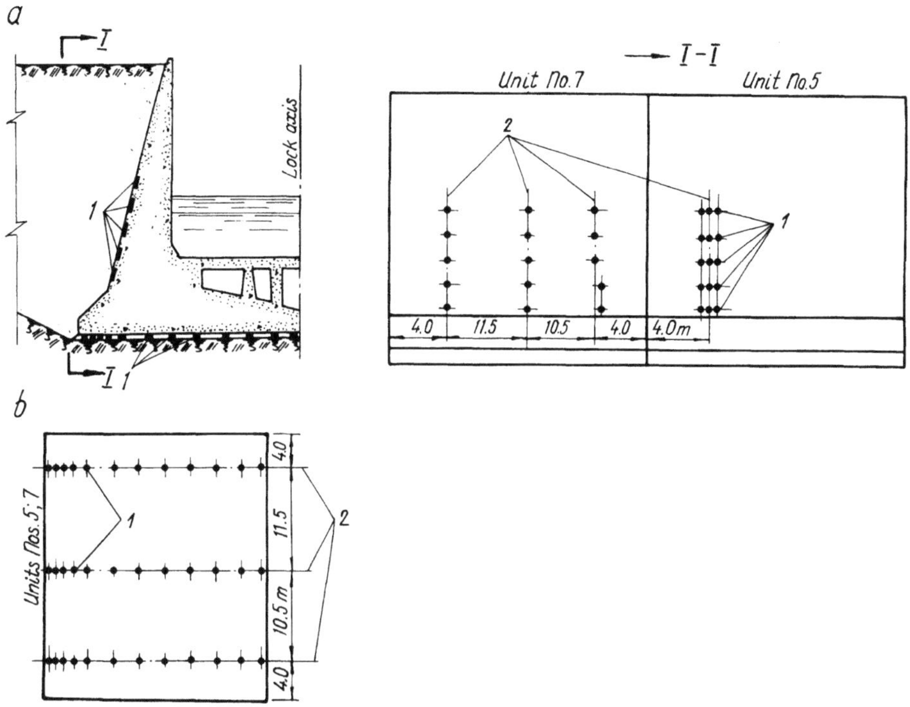

Figure 6-4. Arrangement of soil pressure cells in a reinforced concrete structure of navigation lock (from Burmistrov, 1961). (a) Typical cross section of chamber of navigation lock, (b) arrangement of soil pressure measuring instruments; 1. Soil pressure cell; 2. locations of instruments.

carried out a number of tests to estimate the errors in cell readings when its active face does not sit flush with the interface. In this case the effects of the magnitude of soil pressure, soil modulus of deformation, and size of the soil pressure bearing membrane (diameter) were also investigated. The performance of standard contact pressure cells installed into a concrete cylindrical pile (d = 530 mm) was also tested. In these experiments the cells were tested in calibration chamber SCC-475-0.5 having a stiff (concrete) insert (for a description of this chamber, see Chapter 3). The load was applied in increments of 0.05 and 0.1 MPa in a range from 0 to 0.5 MPa.

Two types of soil pressure cells, namely the one depicted in Figure 4-1 (d = 50 mm) and another SDKS-type cell (d = 135 mm) with a hydraulic multiplicator (Figure 4-8), were used in this investigation. Both instruments were able to be raised 2.5, 5, 7, and 14 mm above the insert's surface by means of special well-polished shims; they could also be dropped below the insert's surface by 2.5 and 4.5 mm. Cell readings during absolutely flush conditions (p_m) were taken as 100%

(reference). The relative error (Δ_p) in actual cell readings (p) attributed to "not absolutely" flush conditions were calculated as

$$\Delta_p = \frac{p_m - p}{p} 100\% \tag{6-1}$$

In the process of soil loading within the calibration chamber the modulus of deformation, E_s, changed at rates as described in Chapter 5, Table 5-4.

The mean values of deviation (error, Δ_p) in soil pressure readings taken by four smaller contact pressure cells at pressures equal to 0.3 and 0.5 MPa were as follows: when the soil pressure-bearing membrane was projected above the interface by 2.5 and 4.5 mm, the mean value of Δ_p was, respectively, equal to 14 and 26.5%, and when the membrane was located 2.5 and 4.5 mm below the interface, the mean value of Δ_p was, respectively, 12 and 13%. Furthermore, it was found that quite substantial changes in the value of the modulus of soil deformation, e.g., E_s = 18.2 MPa at p = 0.3 MPa during the first cycle of cell loading and E_s = 108 MPa at equal loading after the second loading cycle, that is approximately six times greater E_s during the second loading cycle, did not significantly affect cell performance.

The error, attributed to changes in E_s, did exceed approximately 2%. It should be noted that the comparatively small active contact area of the instruments in question resulted in a substantial scatter in cell readings when the applied pressure was below 0.3 MPa. This did not permit definite conclusions to be drawn on errors in cell readings under $p < 0.3$ MPa. This was covered by tests carried out on three SDKS-type pressure cells having a much larger contact area. In these tests the cell was projected above the interface by 2.5, 5, 7, and 14 mm; the results are shown in Figure 6-5. From this figure it is seen that the error in cell readings grows faster with an increase in loading in the first cycle than in the second cycle of loading.

Similar to previously described tests, the effect of a change in E_s did not have significant effects on the value of the error in cell readings. A comparison of tests carried out on both types of cells presented in the form of Δ_p (%) = $f(\delta/d)$ is given in Figure 6-6. Here δ is the projection of the cell's soil pressure-bearing face above or below the interface and d is the cell diameter. The upper portion of the Δ_p (%) = $f(\delta/d)$ diagram represents results of tests in which the cell membrane was projected above the interface, and the bottom part illustrates Δ_p (%) = $f(\delta/d)$ for the case in which the membrane was displaced below the interface. From this figure we can conclude that the relationship between Δ_p and δ/d is practically linear.

Smaller cells were also installed into a convex concrete surface that simulated the surface of a concrete pile 530 mm in diameter. The cells, loaded by pressures equal to 0.3, 0.4, and 0.5 MPa, were installed "flush" and 2.5 and 5 mm above

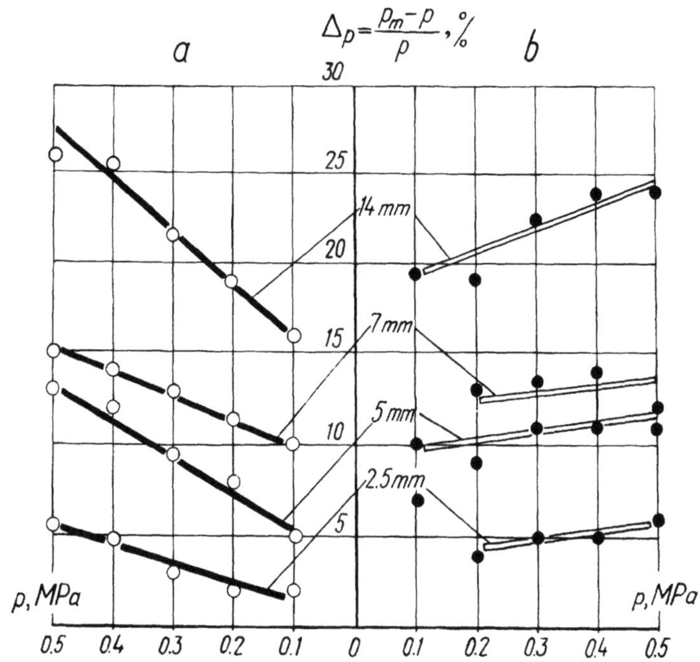

Figure 6-5. Relationship between relative error in soil pressure readings, Δ_p, taken by the SDKS-type pressure cell and distance between the cell active surface and soil/insert interface, e.g., 2.5, 5, 7, and 14 mm. (a) First loading cycle; (b) second loading cycle.

the concrete surface. In this case the minimum averaged value of the error of the cell readings, Δ_p, was, respectively, equal to approximately 6, 12.5, and 31%.

The conclusions drawn from the described experiments are as follows.

- The pressure contact cells must be installed flush with the interface as close to the absolute conditions as practical.
- For instruments that are not installed perfectly flush and deviate from absolute, appropriate corrections to the cell readings should be considered. The "expected" magnitudes of these corrections can be obtained from the data presented in Figure 6-6.

6.3 Influence of Size of Soil Grains on Cell Performance

Review of the existing literature and the available reports by this writer suggest that to date this subject has not been given proper attention. In granular media the pressure on the active surface of the pressure cell is transmitted through the

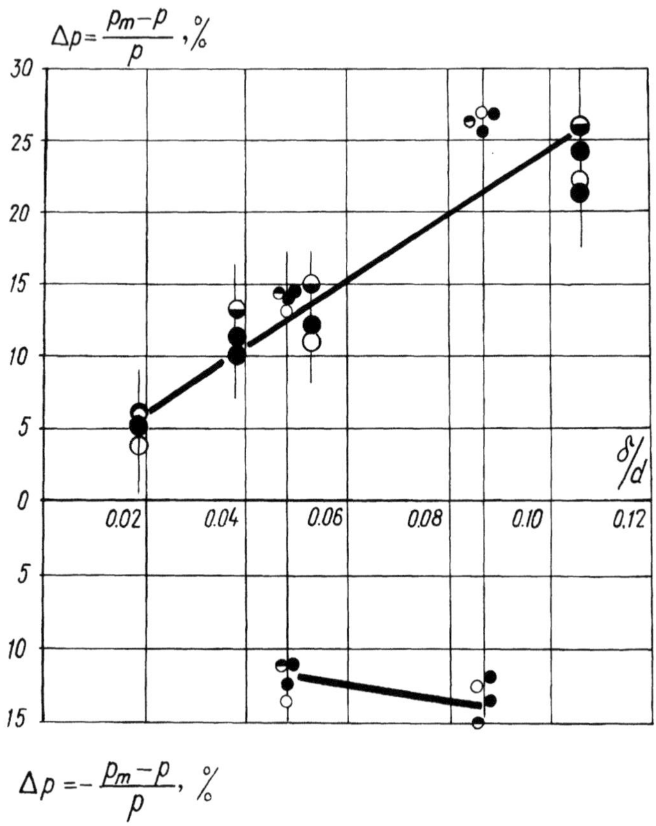

Figure 6-6. Relationship between the relative error in pressure cell performance and the ratio of misalignment of cell soil pressure-bearing surface, δ, and cell diameter, *d*. *Note*: Smaller circles correspond to cells having smaller *d*; larger circles correspond to cells with larger *d*. Light circles correspond to first loading cycle, black ones correspond to second loading cycle.

soil grains. Accordingly, the number of these points is smaller in coarse grain sands and larger in fine sands. The number of these points and the intensity of the load that is transmitted via these points to the cell may change as a result of increased pressure. This happens because of displacement and change in orientation of the sand grains. This, in turn, may affect the soil pressure readings taken from the cell.

In order to standardize the investigation process in the former USSR, it was stipulated that the "standard" granular media used in all related experiments should be an air-dry quartz sand having grain sizes in a range of 0.25 to 0.50 mm. Kallstenius and Bergau (1956) attempted a theoretical evaluation of the influence of grain size on the error in soil pressure measurements conducted with contact pressure cells having a stiff membrane displaced plane-parallel under the effects of pressure. They modeled a granular media as a sand having spherical grains of the same diameter; it was assumed that at all load conditions the contact point pressure of all grains was the same.

Based on these assumptions, these investigators concluded that if the expected error in soil pressure measurements attributed to grain size should be limited by 3%, then the diameter of the grains should not exceed 2% of the cell's active surface diameter. In reality, however, sands have grains of different sizes. For example, in most coarse-grain sands, particles in a range of 1 to 0.5 mm constitute approximately 85% of the mass, and in medium sands particles ranging from 0.5 to 0.05 mm constitute approximately 90% of the sand mass. Hence, in the absence of credible experimental data, it is possible to conclude that instruments having membrane diameters $d \geq 50$ mm (from the grain size point of view) can be used with confidence for soil pressure measurement. However, in general, the larger the diameter of the cell's active surface, the better the expected results can be.

6.4 Interaction Between Adjacent Pressure Cells in a Mass of Soil

In order to obtain reliable readings from soil pressure cells installed in a mass of soil, these instruments must not affect each other by being too close in proximity. This applies to both the vertical and horizontal placements of soil pressure cells in the soil mass. On the basis of theoretical investigations on the effects of hard elements placed within a granular media, Sammal (1980) concluded that under these conditions stress concentrations (a stress field) around these elements should be expected. Sammal also arrived at the conclusion that the maximum additional vertical stress, $\Delta\sigma_z$, in the vicinity of the hard element that is embedded within the granular media does not exceed 10% of the outside stress, σ_z, applied to the granular media at a distance $z \geq 0.2r$, where r is the radius of the hard element, assuming it is circular.

Furthermore, the horizontal stresses at the edges of the body of the hard element are basically insignificant (less than 2%), and in this sense the two closely located hard bodies do not affect each other. On the basis of photoelastic and theoretical investigations, Sammal (1980) concluded that in practical applications, e.g., investigation of soil/structure interaction, structure behavior monitoring programs, etc., in order to avoid adverse effects upon each other, the soil pressure measuring instruments should be located vertically in the soil mass at a distance equal to and no less than $3d$ between contact surfaces, and equal to d_1 between edges of the adjacent instruments in a horizontal plane. Here d is the diameter of the instrument's active surface.

Sammal's work is based on the premise that a granular medium is elastic. This, however, is not the case in real soils. In the past three decades several important investigations have been carried out in the former USSR aimed at perfecting the soil pressure measuring instruments and developing guidelines for instrument installation and use.

Baranov et al. (1971a, b) concluded that stiff flat cells produce no adverse effect upon each other in a horizontal plane when placed at distances 20 mm or more from each other. In a vertical plane at a distance equal to $1.5d_1$ between cells, the upper cell affects readings by the lower one, but the lower instrument does not affect the performance of the upper one. Cells have no effect on each other when the distance between then is equal to $3d_1$. Sivolap and Ivanov (1977) concluded that in the vertical plane soil pressure cells can be installed even closer, e.g., at distance $2d_1$ with no effects upon each other. When the same instruments were tested side by side, it was found that the error in soil pressure measurements by the adjacent cells in dense sand did not exceed 3.5%, and in loose sand the error reached 8.5%. Sivolap and Ivanov found that at a distance of $1.5d_1$ between edges of adjacent stiff flat cells, the cells had virtually no effect upon each other.

These experiments also indicated that nonstiff pressure cells closely located in a vertical plane ($< 2d$) produced substantial effects upon each other. For example, the errors in readings taken from the lower instrument could reach up to 40%; the interaction between the upper and lower instruments was practically nonexistent at a distance between them equal to 4 to $5d_1$. Some practical recommendations that stem from the above and other reported results of cell investigations are as follows.

- The effects of the adjacent pressure cells installed in a granular medium depend heavily on the soil density and the stiffness of the instruments.
- When $E_{epc}/E_s \geq 10$, the distance between adjacent cells in the vertical plane should not be less than 2.5 to $3d_1$, whereas the space between instruments placed side by side should not be less than 1 to $1.5d_1$.
- When $E_{epc}/E_c < 10$, the above distances, respectively, should not be less than 5 to $6d$ and $2d_1$. Here E_{epc} is the modulus of elasticity of the soil pressure cells, E_s is the soil modulus of deformation, and d is the instrument diameter. In practical applications, most soil pressure measuring instruments installed in a mass of soil for protection from mechanical damages are furnished with a rubber ring. Examples are found elsewhere in this work.

As pointed out by Sammal (1980), the stress concentration around the pressure cell is higher in the zone of decreased stiffnesses, i.e., in the case in question in the area of the rubber ring that frames the instrument's edge. Where relevant, the rubber ring also serves as a damping interlayer, which has a potential to substantially reduce the transfer of horizontal forces to the instrument's body.

This writer and co-workers investigated the effects of the rubber ring on the performance of a stiff flat soil pressure cell having a vibrating wire type transducer. The experiments were conducted on two similar instruments, one did not have the rubber ring, while the other had a ring. The investigation was carried out in loose and dense sand in which the cells were loaded by pressures equal to 0.2 and 0.3 MPa. Readings from cells without the rubber rings were taking as 100% (reference). It was found that the average reading from six tests with the cell having a rubber ring was approximately 5.4% higher than that in similar tests on a cell without the rubber ring. This most likely occurs because the rubber ring reduces the horizontal stresses on the instrument when it is loaded in a granular medium. This, in turn, leads to concentration of pressures in the region of soil adjoining the upper and lower contact surfaces of the pressure cell, and the cell measures these increased pressures. This assumption is supported by test results discussed in Chapter 2. Thus, we can state that, in general, the framing rubber ring increases the sensitivity of the stiff flat soil pressure cell.

7

Contact Stresses at Interface Between a Stiff Foundation Block and Foundation Soil: Large-Scale Model Tests

Large-scale model tests that study the contact stresses at the interface of a stiff foundation block and bearing soil are discussed in detail by Lazebnik and Smirnov (1964), Lazebnik et al. (1966), and Lazebnik (1966). The contact stresses were investigated using stiff concrete blocks of various shapes, e.g., round slabs having diameters 1.1 m and 2.2 m and a rectangular block 3 × 1 m in plan. This investigation was preceded by a series of preliminary experiments aimed at developing the proper investigation procedures and selecting the proper pressure measuring instruments. In these preliminary experiments a reinforced concrete block 1.8 × 0.6 m, and 0.7 m deep, was tested on foundation soils represented by sand and loam. Loads up to 0.4 MPa were applied by means of a hydraulic jack in increments of 0.1 MPa for sand and 0.05 MPa for loam. Each test consisted of two subsequent loading cycles. The contact soil pressure was measured by means of instruments having pressure measuring systems, e.g., bonded wire strain gages and vibrating wire transducers.

The soil pressure was measured directly at the block/soil interface and at a certain distance from the block's underside. This resulted in soil pressure diagrams, which were verified against the applied loads. Based on the required equilibrium conditions between the applied load and instrument readings, the best experiment procedure was selected and used in the follow-up investigations. These preliminary tests indicated that the best measurement results for contact soil pressure can be achieved by using stiff pressure cells installed perfectly flush with the underside of the footing.

The shape of the soil reaction diagram also provided the investigators with good ideas on part of the arrangement (layout) of the pressure measuring instruments. The follow-up experiments, with larger stiff concrete blocks of different configurations, were carried out in a thick-walled concrete bin 30 × 5.5 m in plan and 5 m deep, filled with a clean, dry quartz river sand. The sand's minimum and maximum densities were, respectively, 15.4 and 18.1 kN/m^3. At a sand density of 16.5 kN/m^3 its angle of internal friction, ϕ, was approximately 31.5°, and at a water content w = 5.2 to 6.1% of the sand's cohesion value, c, was equal to approximately 0.008 MPa.

During the experiments the water content, w, varied within the range ± 1% to ± 3%. Sand placement was carried out in layers with follow-up densification by a platform-type vibrator to a density required by the test program. The soil density was carefully controlled. The test blocks were installed on a thoroughly leveled surface that ensured good contact between the block and foundation soil. As noted earlier, the test load was applied in increments of 0.05 and 0.1 MPa. Each load step was followed by measurement of the contact pressure at the footing/soil interface and measurement of the associated settlement of the block.

All test blocks were loaded twice and in some instances three times in order to investigate the effects of cyclic loading as well as unloading on changes in the footing/soil interaction. Finally, it should be noted that in all experiments stiff instruments of SDKS type with vibrating wire transducing systems were used for the measurement of contact soil pressures.

7.1 Round Stiff Concrete Block 1.1 m in Diameter

The soil pressure measuring instruments' arrangement in this block and its general view are shown in Figure 2-8. The instruments are located, respectively, 275 and 470 mm from the block's center. The results of the most representative tests are shown in Figure 7-1. From this figure it is seen that at the first loading cycle the maximum contact stress is at a distance 275 mm from the center of the block; in the second loading cycle the maximum stress area has shifted to a distance 470 mm from the center. Note that the results of the contact stresses presented in Figure 7-1 are average readings taken from six identically loaded pressure measuring cells.

Again, changes in the soil pressure diagram in the second loading cycle are attributed to substantial changes in soil density as a result of the first cycle of loading. For example, before the foundation soil was loaded, its relative density was D_r = 40.9% (first test) and D_r = 52.2%. Accordingly, the initial soil's unit weight of 16.4 kN/m^3 changed to 16.7 kN/m^3. Changes in stresses at the block's center, intermediate area, and at the edges as well as block settlement during the first and second loading cycles are illustrated in Figure 7-2. The associated changes in the modulus of deformation of the foundation soil are depicted in Figure 7-3.

The data presented in these figures underline the effects of changes in soil deformation characteristics on the shape of the soil reaction diagram, values of the contact stresses, and settlement of a stiff footing. This was confirmed by subsequent experiments carried out on blocks of other sizes and shapes.

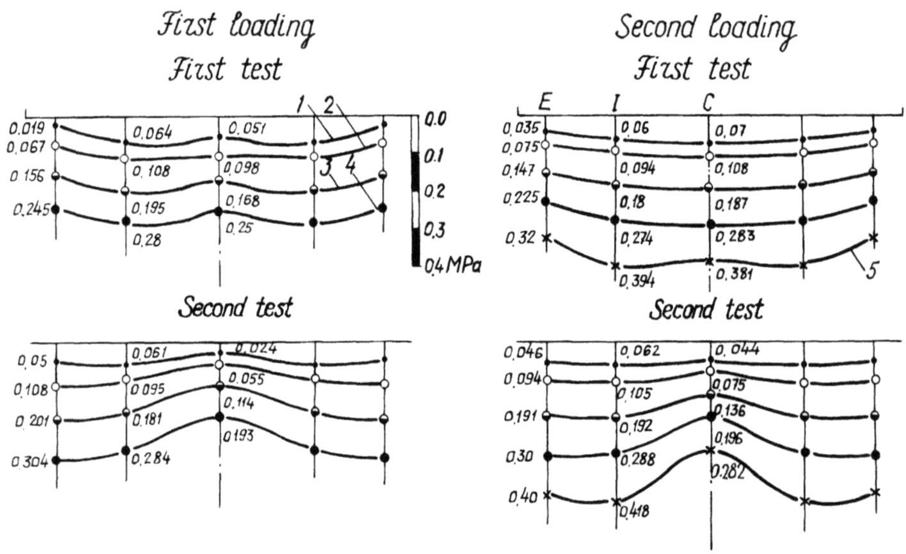

Figure 7-1. Distribution of soil contact pressure under round stiff slab 1.1 m in diameter. 1 through 5. Soil reaction when block is loaded respectively by 0.05, 0.1, 0.2, 0.3, and 0.4 MPa.

7.2 Round Stiff Concrete Block 2.2 m in Diameter

The arrangement of the contact soil pressure cells in this set of tests is shown in Figure 7-4. The foundation soil conditions in these experiments were almost identical to those of the previously described tests. The test results are illustrated in Figures 7-5 through 7-7. From these figures it follows that, similar to the experiments conducted on the 1.1-m-diameter block, the contact stresses, σ_i, increase almost linearly with an increase in application load, p; i.e., σ_i is proportional to p. These experiments also confirmed substantial changes in the soil modulus of deformation as a result of the application of compressive loads.

7.3 Stiff Rectangular Concrete Block 3 × 1 × 0.6 m

This block and its layout are illustrated in Figures 7-8 and 7-9. Five rows of pressure cells, with five cells in each row, were installed in this block to investigate soil reaction stresses at the underside of the block in response to external pressure. All tests were repeated three times; the average results of the soil pressure measurements are given in Figures 7-10 and 7-11.

As shown in Figure 7-10, the soil reaction diagram at the central cross section 3-3 has a concave shape, which is characteristic for all load increments in both loading cycles. This confirms the test results obtained in the two previously

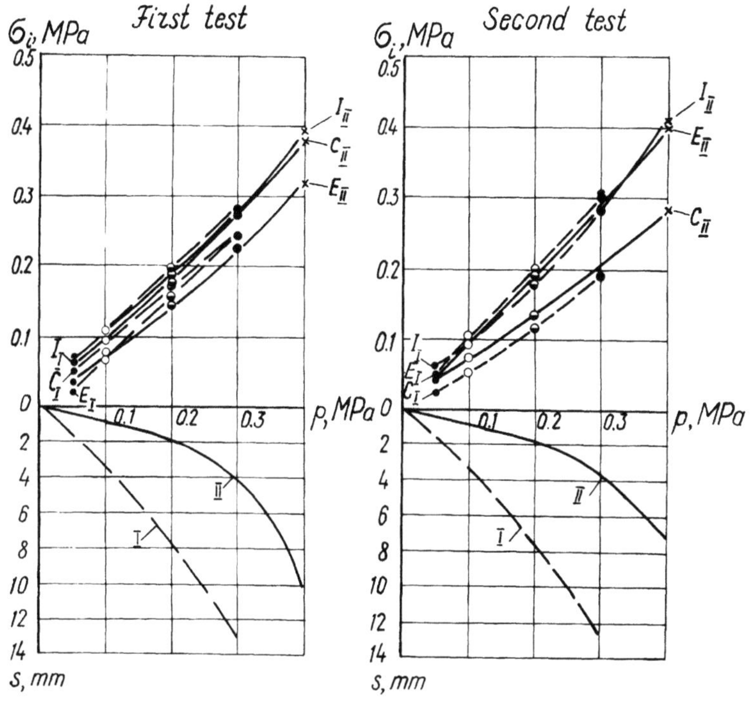

Figure 7-2. Round stiff block 1.1 m in diameter. Averaged contact soil pressure ordinates, σ_i, at block center (*C*), intermediate area (*I*), and edge (*E*), and block settlement, *s*, at first (I) and second (II) loading cycles.

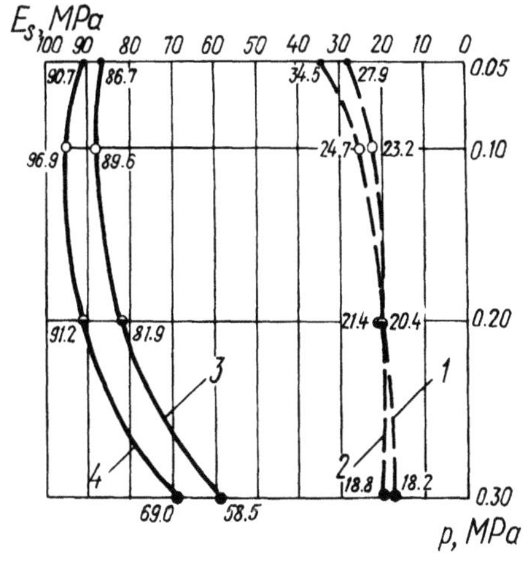

Figure 7-3. Changes in slab modulus of deformation, E_s, with load increase on round concrete block 1.1 m in diameter. 1 and 3. In the first test; 2 and 4. in the second test.

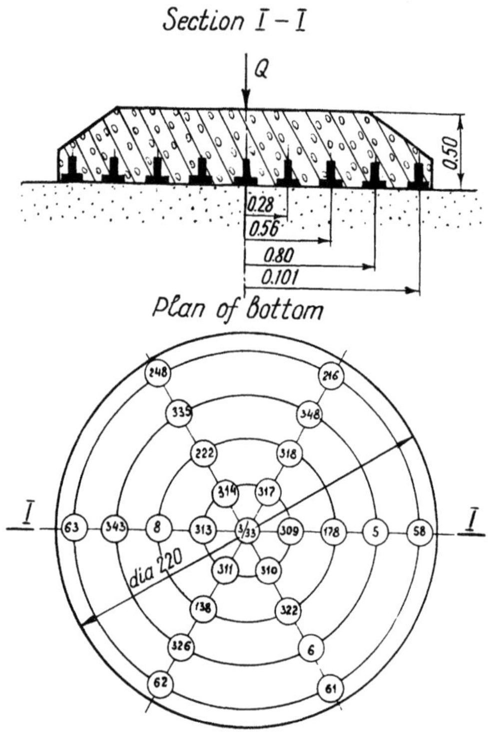

Figure 7-4. Round concrete test block 2.2 m in diameter. Layout of contact soil pressure measuring instruments.

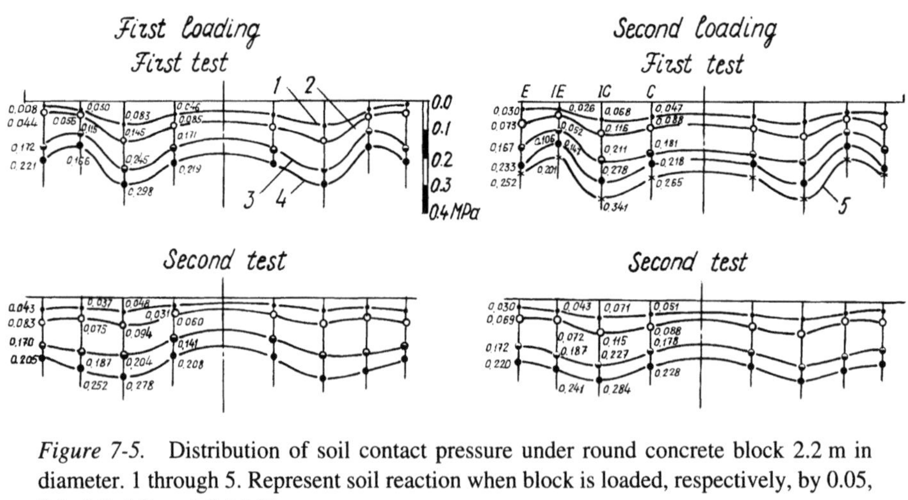

Figure 7-5. Distribution of soil contact pressure under round concrete block 2.2 m in diameter. 1 through 5. Represent soil reaction when block is loaded, respectively, by 0.05, 0.1, 0.2, 0.3, and 0.4 MPa.

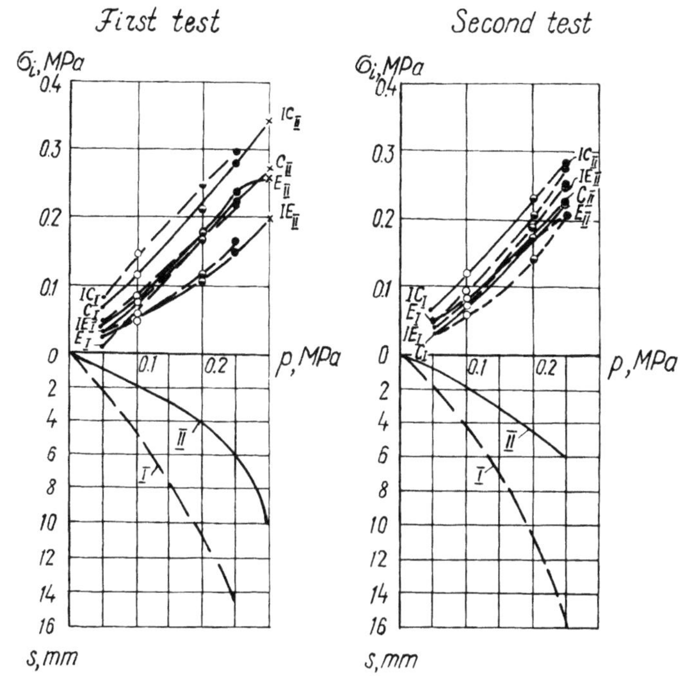

Figure 7-6. Round concrete block 2.2 m in diameter. Averaged contact pressure ordinates, σ_i, and block settlement, s, at first (I) and second (II) loading cycles.

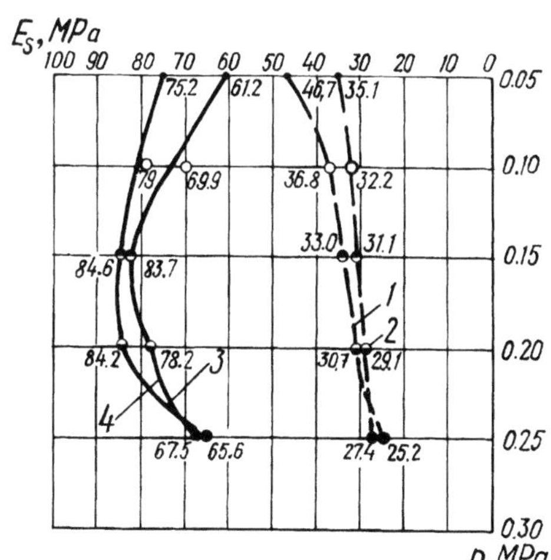

Figure 7-7. Changes in soil modulus of deformation, E_s, with load increase on concrete block 2.2 m in diameter. 1 and 3. In the first test; 2 and 4. in the second test.

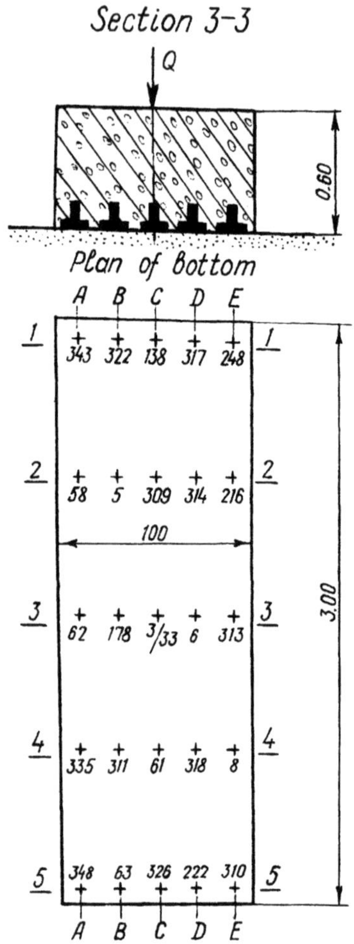

Figure 7-8. Concrete rectangular block 3 × 1 × 0.6 m and layout of contact soil pressure measuring instruments installed flush with block underside.

described case histories. The shape of the soil pressure diagrams at intermediate sections 2-2 and 4-4 at all loading stages and cycles is convex, and at the end sections 1-1 and 5-5 the shape of the pressure diagram looks like a mirror image of the diagram at section 3-3.

Such complicated stress distributions are explained by stress concentrations at the block corners and edges and the development of plasticity zones in these areas. Longitudinally (Figure 7-11) the shapes of the pressure diagrams at the block edges and center are essentially identical. The shape of the intermediate diagrams is basically convex. As seen in Figures 7-10 and 7-11, at the second loading cycle the edge stresses are slightly higher than those resulting from the

Figure 7-9. General view of 3 × 1 × 0.6 m concrete rectangular block. Process of installation of soil pressure cells.

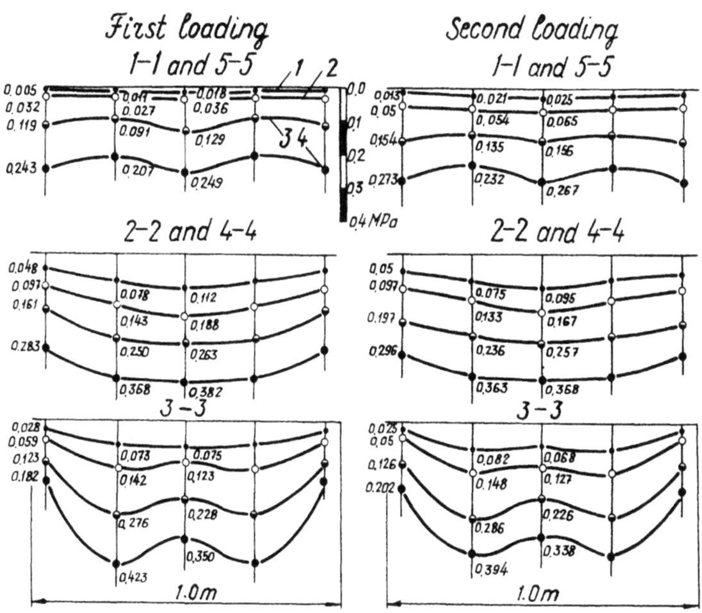

Figure 7-10. Distribution of soil contact pressures at interface of 3 × 1 × 0.6 m stiff concrete block and foundation soil. 1 through 4. Measured contact soil pressure diagrams at cross sections as indicated in Figure 7-8; load increments are, respectively, 0.05, 0.1, 0.2, and 0.3 MPa.

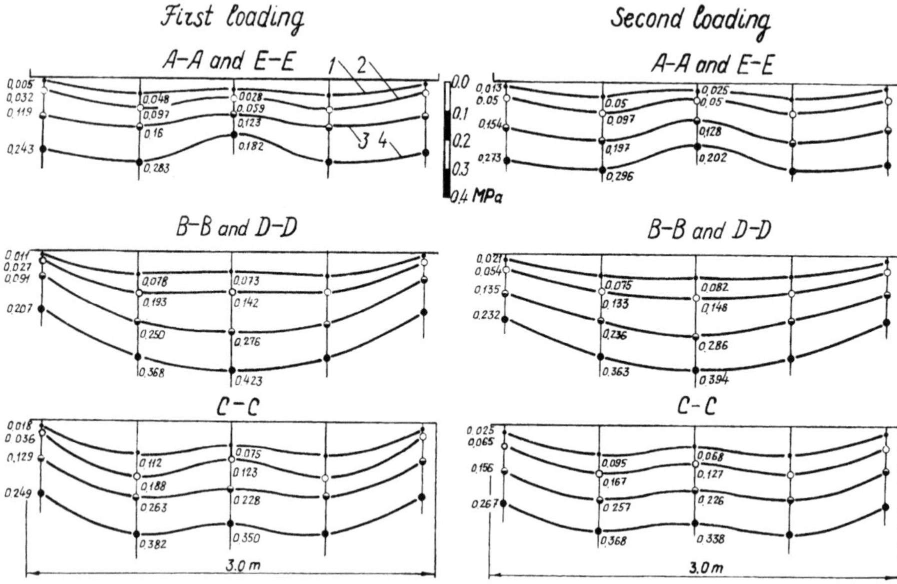

Figure 7-11. Distribution of soil contact pressures in longitudinal direction at interface of 3 × 1 × 0.6 m concrete block and foundation soil. 1 through 4. Contact soil pressure diagrams; load increments are, respectively, 0.05, 0.1, 0.2, and 0.3 MPa. *Note*: This distribution is based on data presented in Figure 7-10.

first loading cycle. The block settlement, s, and changes in the modulus of deformation of the foundation soil, E_s, are similar to those obtained in the previously described tests with round blocks.

The conclusion that can be drawn from all three test series is that during the second cycle of block loading stress concentrations at the block edges increase. This, in turn, results in increased bending moments in the footing.

8

Field Investigation. Case Histories

This chapter contains descriptions of some characteristic field investigations (monitoring programs) conducted by this writer with co-workers during the period from 1965 to 1991.

8.1 Round Concrete Foundation for Cylindrical Concrete Shaft to Support a Cable Roof System

This foundation system is part of the large bus parking station constructed in Kiev, Ukraine. A general view of this structure is shown in Figure 8-1. It includes a cylindrical reinforced concrete shaft supported on a round, very stiff foundation system comprised of a heavily reinforced concrete slab 0.8 m thick and 24.6 m in diameter, eight radial stiffening walls 0.9 m thick and 4.2 m deep, and a top reinforced concrete slab 0.3 m thick. A cross section of this structure is depicted in Figure 8-2. The foundation is embedded 5 m below grade and founded upon very fine sands underlain by fine and clayey sands. Soil deformation parameters were determined via in situ tests at depths of 5 and 7.5 m. The deformation modulus of the founding soil is approximately equal to 12 MPa; that of the underlying soils is approximately equal to 13 MPa.

The foundation system is designed to support the weight of the structure itself and the substantial snow load that collects on the parking station roof. The foundation structure system is very stiff and, according to flexibility criteria defined by Gorbunov-Possadov and Malikova (1973) as approximately equal to $10(E_s d^3)/(E_f h^3)$, it may be classified as "absolutely stiff." Here E_s and E_f are, respectively, moduli of deformation of the foundation soil and the footing, and d and h are the foundation diameter and its "equivalent" thickness.

Sixteen soil pressure measuring cells of SDKS type with vibrating wire transducing systems were installed flush at the underside of the foundation slab. The

Figure 8-1. Bus parking terminal in Kiev, Ukraine under construction. At this stage the central shaft is completed and the roof cables are installed and functional.

layout of these instruments is shown in plan view of Figure 8-2. In order to obtain a "flush" condition as accurate as possible, all instruments were initially installed in special concrete blocks where they were fixed flush with the concrete surface. These blocks were then placed on the carefully leveled foundation soil at predetermined locations and fixed in this position by carefully placing a 100-mm-thick layer of concrete around the blocks. This was followed by installation of structural reinforcing steel and then the casting of concrete for the foundation slab.

The soil pressure readings were taking at the end of presubscribed construction stages as indicated in Figure 8-2. At each of these stages the estimated active loads were compared with those measured. Comparison of both shows some divergence in the values of the acting and measured reacting forces; at the start of construction it was in the neighborhood of 8–10% and towards the end it became larger; however, it did not exceed 12.4%. From the soil reaction diagrams shown in Figure 8-2, it follows that at the initial construction phase (diagram 1) when the load actually comes from the flat slab and the stabilizing blocks installed upon this slab, the soil reaction diagram is concave with the maximum stresses located at the slab's outer perimeter. All following soil pressure diagrams that resulted from the added structural weights of foundation and roof systems as well as from the snow load are concave-convex. The shape of these diagrams actually reflects the structural peculiarities of the footing/shaft system; they are convex at the stiffer part located at the perimeter of the cylindrical shaft and concave where the slab inward deflections are expected, e.g., at the slab outer

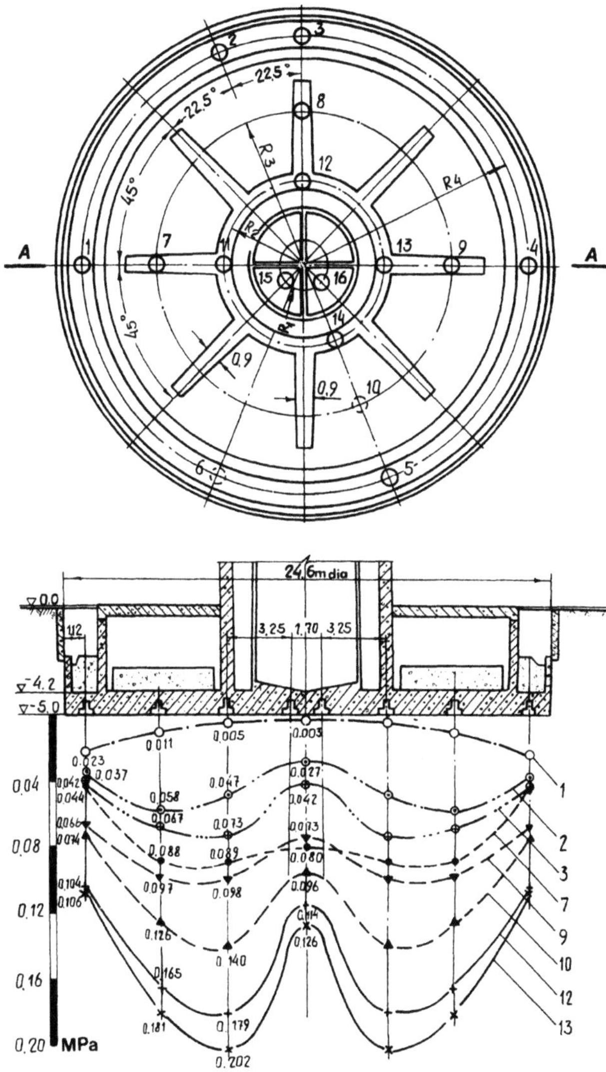

Figure 8-2. Bus parking terminal in Kiev, Ukraine. Plan and typical cross section of the roof-supporting foundation and soil reaction diagrams at interface of foundation slab and foundation soil related to miscellaneous construction phases. 1. Foundation slab is completed, outer wall is erected, and stabilizing concrete blocks are cast; 2. radial cells and outer ring wall are completed, and structure is backfilled with granular soil; 3. foundation structure cover slab is installed and outer concrete ring wall is completed; 7. central shaft is completed, and roof cables are installed and tensioned; installation of concrete roof slabs just begun; 9. eight radial rows of roof slabs are installed; 10. twelve radial rows of roof slabs are installed; 12. all roof slabs are installed; 13. completed roof is loaded with snow (approximately 1000 tonnes).

perimeter and its central part were there are no stiffening ribs. The slab deflection at the foundation center is approximately 5 mm.

8.2 Cellular Square Footing Constructed on Granular Foundation Soils

A typical cross section of this structure is illustrated in Figure 8-3. It has been constructed to support a section of the Energodar, Ukraine, nuclear power station that houses a nuclear reactor. The total weight of the system, including the footing, reactor, and powerhouse building is approximately 250,000 tonnes. The reactor building is circular in cross section with an outside diameter equal to 47.4 m. The cellular footing structure is 68 m × 68 m in plan and 19.8 m deep. The underside of this structure is located 6.6 m below grade. By all definitions this structure is considered to be very stiff.

The reactor building is constructed upon a stratum of medium to fine river sand of medium density ($\phi = 32°$, $c = 0.001$ MPa); the depth of this stratum is approximately 28 m. This is underlain by the following soils: sandy loam and loam ($\phi = 25°$, $c = 0.019$ MPa, $h = 18$ m); very fine sand ($\phi = 24°$, $c = 0$, $h = 1$ m); sandstone ($\phi = 24°$, $c = 0.08$ MPa, $h = 1.5$ m); kaolinite ($\phi = 23°$, $c = 0.027$ MPa, $h = 18.5$ m); and bedrock. Here ϕ is the angle of internal friction, c is the cohesion and h is the depth of the respective soil stratum.

The mean load (q_1) upon the foundation soil from the weight of the footing is equal to 0.22 MPa. The 1.2-m-thick concrete circular-walled reactor housing, along with the associated power generating equipment, produces a mean pressure

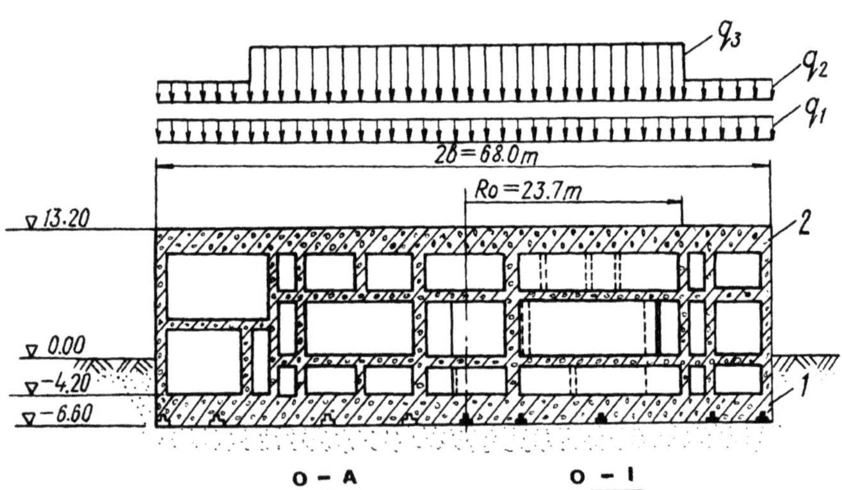

Figure 8-3. Energodar nuclear power station, Ukraine. Cellular square footing to support nuclear reactor. Typical cross section. 1. Foundation slab; 2. reactor floor. *Note*: Foundation plan is given in Figure 8-5.

of $q_3 \simeq 0.511$ MPa, and miscellaneous loads located outside the building structure produce a mean load (q_2) of approximately 0.203 MPa. Because the structure is symmetrical, the resultant load acts through the footing's center. The load distribution is shown in Figure 8-3, and a general view of the reactor block is shown in Figure 8-4.

The contact soil pressures were measured by a total of 36 instruments, the SDKS type, as described in Chapter 4. This was part of a comprehensive effort during which structure settlement, tilt, stresses in the reinforcement, and other relevant parameters of concerning structural behavior were monitored. (Lazebnik et al. 1989) The layout of the contact soil pressure cells is shown in Figure 8-5. The instrument's "flush" installation condition was achieved by using an installation technique similar to that described in the previous case history, namely preinstalling the pressure cells into small concrete blocks (Figure 8-6).

The contact soil pressures measured at different construction stages are depicted in Figure 8-7. The soil reaction diagrams shown in this figure illustrate pressure distributions along axis A-A (left side section O-A) and axis I-I (right side section O-I), as are shown in Figure 8-5. From Figure 8-7 it is seen that from the initial construction stage a load concentration exists at the central part of the footing. The stress concentration in this area intensifies beginning from construction phase 4 and continues through phase 10.

Figure 8-4. Energodar nuclear power station, Ukraine. Power generating unit. General view.

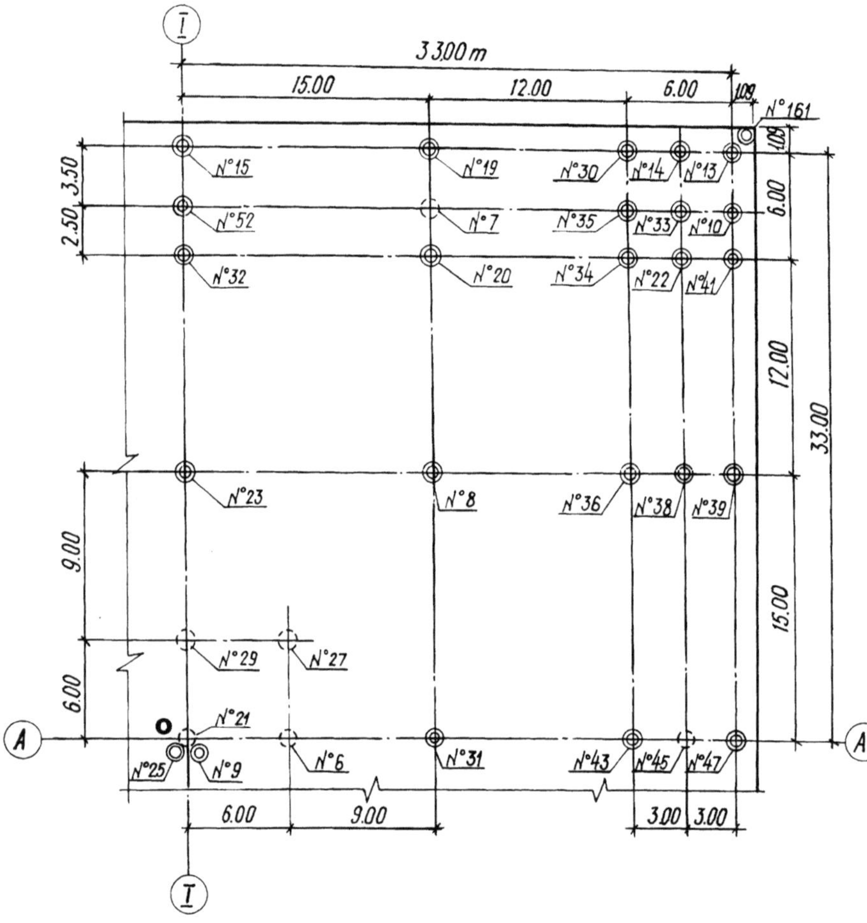

Figure 8-5. Energodar nuclear power station, Ukraine. Layout of contact soil pressure cells in ¼ part of foundation slab.

Contact (edge) stresses at the I-I axis were heavily influenced by construction of the turbine hall building adjacent to the reactor building (see Figure 8-8). This resulted in uneven settlement of the reactor housing structure; the latter has tilted towards the turbine hall at the end of construction by 22.7 mm; the structure average settlement was approximately 85 mm. The increase in edge contact pressures was also affected by soil backfilling procedures. As noted earlier, the soil reaction diagrams shown in Figure 8-7 represent the contact soil pressures at axes A-A and I-I. The diagonal soil reaction diagram and three-dimensional (3-D) maximum soil pressures distribution are illustrated in Figures 8-9 and 8-10, respectively. In general, the differences between the applied loads and measured pressures (accuracy) did not exceed approximately 20%.

Figure 8-6. Concrete blocks with steel nests for installation of soil pressure cells used in field investigations.

8.3 Cellular Square Footing Constructed on Hard Foundation Soil

A nuclear power station, similar to the one described in the previous section, was constructed at Khmelnitsky, Ukraine. In this case the foundation soils were comprised from the top down as follows:

- 8- to 22-m-thick stratum of siltstone-argillite with an average density of 23.4 kN/m^3; density and porosity of dry soil are, respectively, 22 kN/m^3 and 19.7%
- 8- to 10-m-thick stratum of argillite with layers of siltstone and sandstone
- 8- to 9-m-thick stratum of cracked tuff-argillite
- At least 12-m-thick stratum of cracked tuff-sandstone.

The ground water level was at 0.5 to 1.5 m below the ground surface. Uniaxial tests conducted on samples obtained from boreholes indicated the strength of siltstone-argillite material as equal to 4.2 MPa when saturated with water, 4.6 MPa at natural water content, and 19.6 MPa in the dry condition.

The structural monitoring program was very similar to that described in the previous section. Contact soil pressures were measured by using modified versions

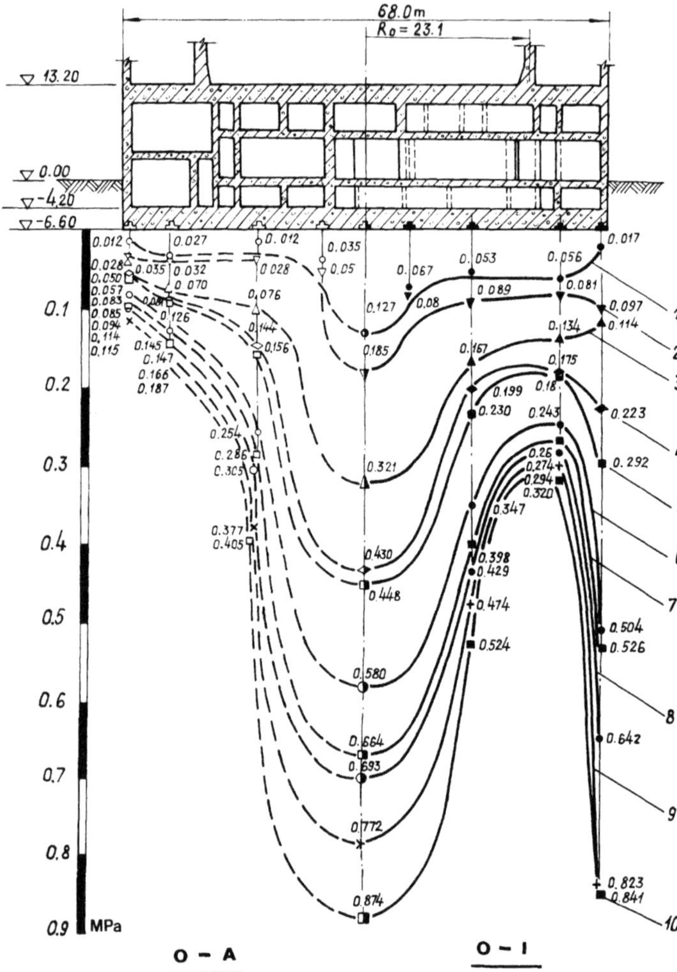

Figure 8-7. Energodar nuclear power station, Ukraine. Contact stress state development in founding soil in sections O-A and O-I in the process of load increase. 1 through 10. Stages of load development. 1. footing is constructed up to grade level, p = 0.1 MPa; 2. footing is constructed to elevation 6.0 m, p = 0.108 MPa; 3. construction has reached elevation 11.0, p = 0.19 MPa; 4. most of upper slab is completed and lower part of concrete shell 7 m high is assembled, p = 0.216 MPa; 5. upper slab and reactor protective shell are completed and some internal structures are erected, p = 0.331 MPa; 6. construction completed up to elevation 20.0 m and some heavy equipment is installed, p = 0.467 MPa; 7. basic part of structure is completed and most of equipment is installed, p = 0.5012 MPa; 8. more weight is added due to block construction and equipment installation, p = 0.51 MPa; 9. powerhouse is completed and most of power generating equipment installed, p = 0.54 MPa; 10. contact soil pressures 3 months after project completion. *Note*: For sections O-A and O-I layout, consult Figure 8-5.

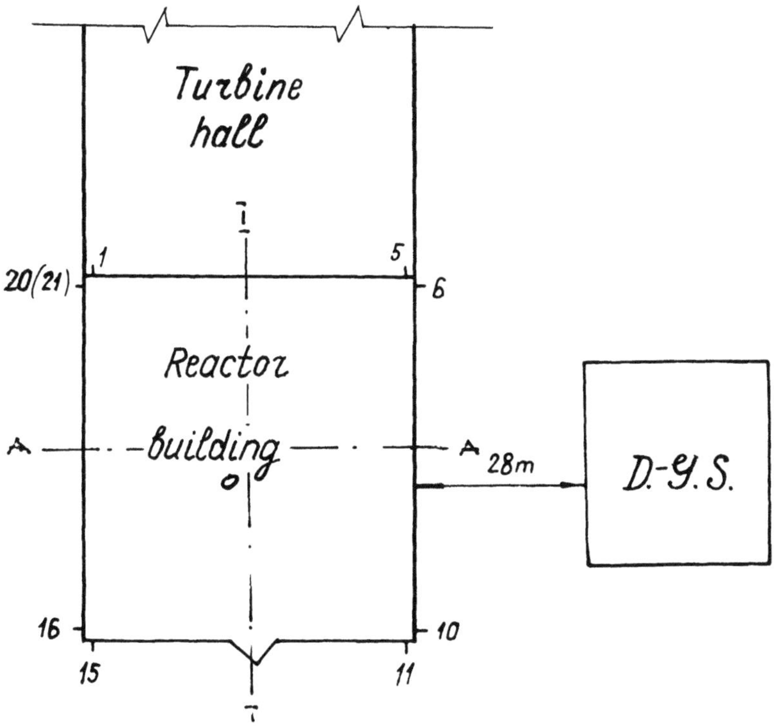

Figure 8-8. Schematic layout of basic structures (power unit N° 4) of Energodar nuclear power station, Ukraine.

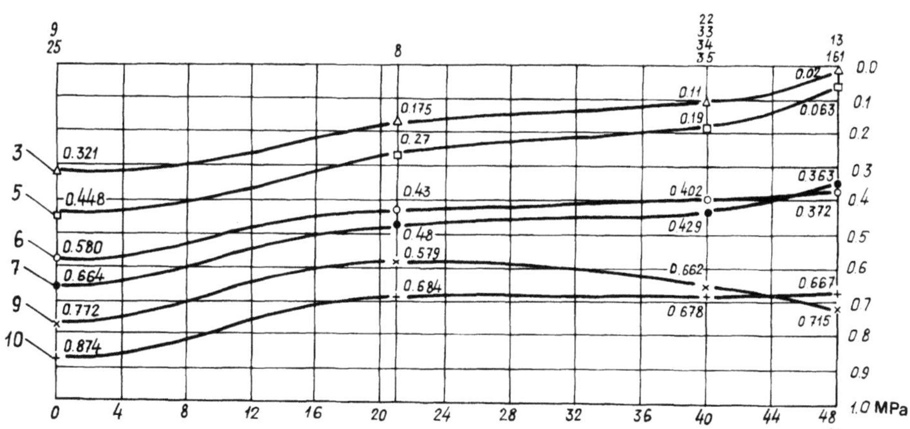

Figure 8-9. Energodar nuclear power station, Ukraine. Diagonal contact soil pressure diagrams. Load phases 3, 5, 6, 7, 9, and 10 are similar to those indicated in Figure 8-7.

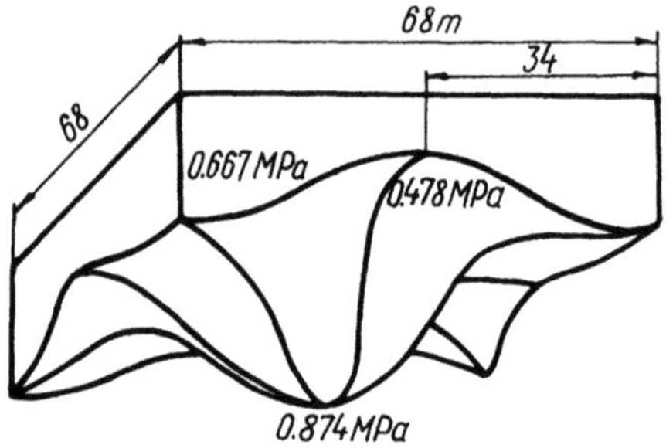

Figure 8-10. Energodar nuclear power station, Ukraine. Three-dimensional representation of distribution of soil reactive pressure.

of the SDKS-type contact pressure cells. This time, however, these instruments were installed symmetrically only at section I-I.

The foundation soil reaction diagrams at different construction stages are shown in Figure 8-11. In this figure every stress point is an average value of readings from two instruments installed symmetrically in section I-I. Because the structure has practically no tilt, the soil reaction diagrams are practically symmetrical. As is seen from Figure 8-11, all contact soil pressure diagrams have consistently increased soil pressure ordinates in the central part of the structure and at its edges, and the stresses have much smaller values in the area located approximately one-quarter of the foundation width from the edges. The obtained values of soil reaction are confirmed by the stress readings within the concrete and in the reinforcing steel.

Comparison of the contact soil pressures distribution for the Energodar and Khmelnitsky nuclear power stations indicates that the shape of the soil reaction diagrams in both cases is essentially similar. Quantitatively, however, the maximum stress value for the Khmelnitsky power station, founded on a more stiff foundation soil, is higher. This conforms to the results of the experiments previously discussed in this chapter. It should be noted that settlement of the Khmelnitsky power station was 34.1% smaller than that of the Energodar power station.

8.4 Foundation of a Tall Building

A 16-storey building was constructed in Kiev, Ukraine (Figure 8-12). Its 38 × 15.5 m foundation slab is comprised of a cast-in-place reinforced concrete slab 0.8 m thick with reinforced concrete ribs projecting 1.1 m above the slab's upper face. A typical cross section and plan view of this slab are shown in Figure 8-13. The structure is founded 11.65 m below grade within a heavy clay stratum with an approximate modulus of deformation $E_s = 60$ MPa. It should be noted that in situ measurements of E_s at specific pressures equal to 250 kN/m^2 produced $E_s \simeq 31$ MPa.

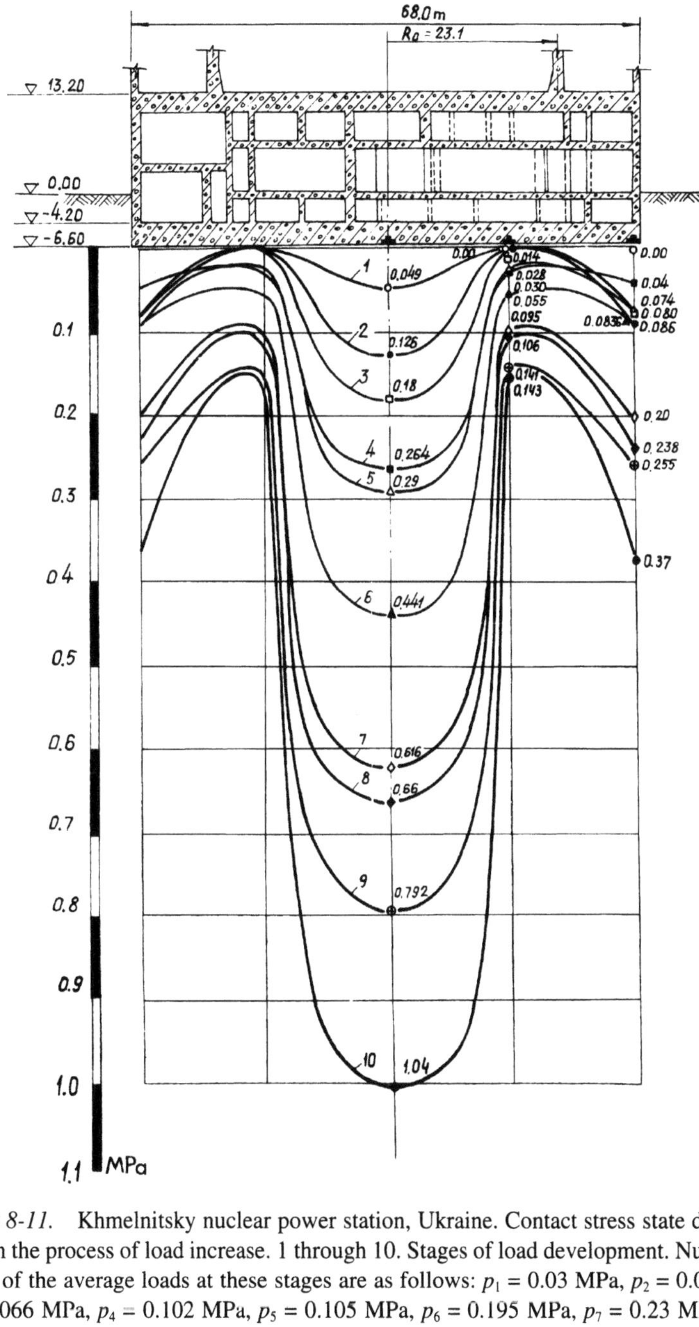

Figure 8-11. Khmelnitsky nuclear power station, Ukraine. Contact stress state development in the process of load increase. 1 through 10. Stages of load development. Numerical values of the average loads at these stages are as follows: p_1 = 0.03 MPa, p_2 = 0.05 MPa, p_3 = 0.066 MPa, p_4 = 0.102 MPa, p_5 = 0.105 MPa, p_6 = 0.195 MPa, p_7 = 0.23 MPa, p_8 = 03 MPa, p_9 = 0.346 MPa, and p_{10} = 0.424 MPa. *Note*: At stage 10 the structure was still incomplete.

Figure 8-12. General view of 16-storey Technical Information Institute building during construction, Kiev, Ukraine.

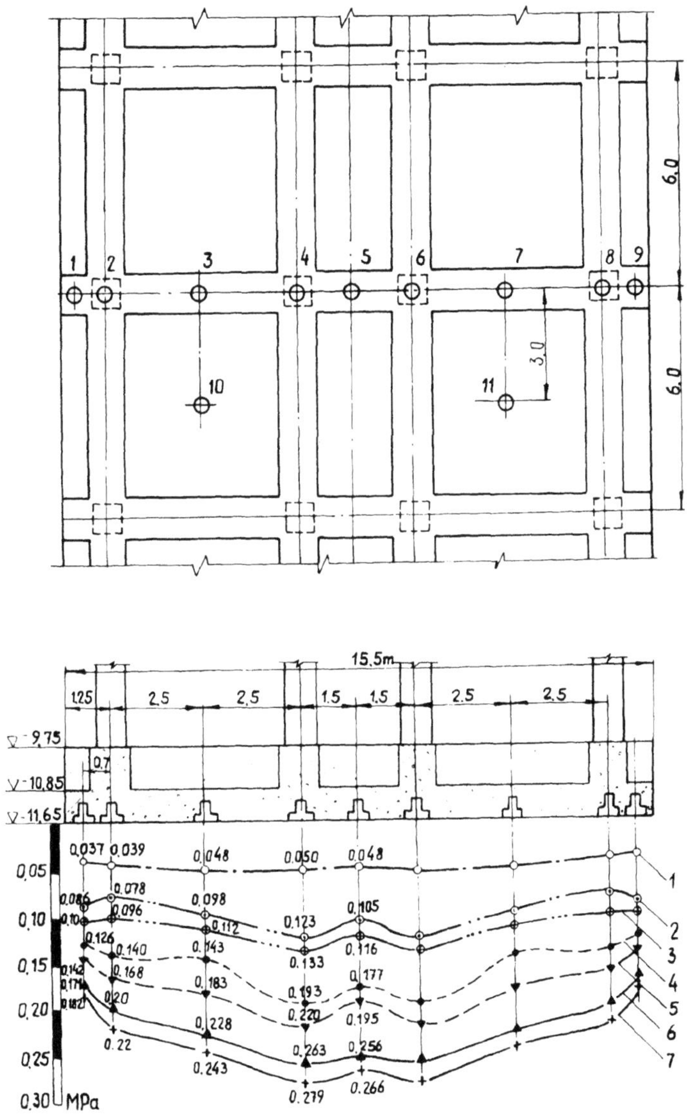

Figure 8-13. Technical Information Institute, Kiev, Ukraine. Plan, typical cross section of foundation slab, layout of soil contact pressure cells, and soil reaction diagrams. 1 through 7. Soil reaction diagrams corresponding to different construction stages as follows: 1. foundation slab and basement columns completed and building equipment installed; 2. three-storey of monolith concrete frame erected; 3. four-storey concrete frame erected; 4. eight-storey concrete frame erected; 5. twelve-storey concrete frame erected; 6. outside and internal panels installed; 7. rest of equipment installed, but finishing work still incomplete.

The measured soil reaction diagrams at various construction stages are shown in Figure 8-13. Soil pressure cells were installed on the founding soil in concrete blocks by using the technique discussed in Section 8.2 (see Figure 8-6). Pressure readings were taken at various construction stages, as is indicated in Figure 8-13. The first set of readings shown in this figure is related to the construction stage when the basement structure of the building was actually completed and the excavation was backfilled with good-quality granular soil. Although, in general, the soil reaction diagram at this construction phase is practically rectangular, some decrease in soil pressure at the foundation edges is quite obvious. Further increases in load on the foundation soil (stage 2) resulted in characteristic changes in the contact soil pressure distribution; these pressures became more concentrated in the area of the centrally located slab-stiffening ribs. Further increases in structural weight followed with increases in soil reaction, particularly in the middle section of the foundation slab (stages 3 through 7). It should be noted that approximately 58% of all structural (and live) loads are carried by the centrally located columns.

The shapes of all soil reaction diagrams are consistent with structure settlement and its local deflections. For example, some decrease in soil pressure at the middle section of the foundation slab is consistent with slab deflection.

8.5 Foundation Slab to Support Loads from Turbine and 300-MW Generator

A concrete slab (Figure 8-14) 1.92 m thick was constructed at the Kostroma thermopower generating station in Russia. This slab was cast upon a layer of lean concrete 20 cm thick and 4 m below grade. The load from the turbine and generator was transmitted to the foundation slab via a network of concrete columns.

The averaged uniformly distributed load upon the foundation soil from the slab structure only is approximately 0.05 MPa; its is approximately equal to 0.145 MPa when loaded by weights of the turbine, generator, and the power equipment supporting structure. The slab is founded on a moraine-type heavy clay with values of $\phi = 18°$, $c = 0.04$ MPa, and $E_s = 30.0$ MPa. The depth of this stratum from grade down is approximately 9.0 m. This is underlain by the following soils:

- heavy loam with inclusions of pebble, $\phi = 26°$, $c = 0.046$ MPa, $E_s = 45$ MPa, with a thickness of 1.5 to 2.7 m
- medium- to fine-grained sand, $\phi = 34°$, $c = 0.002$ MPa, $E_s = 35$ to 60 MPa, with a thickness of 10 to 11 m
- plastic clay, $\phi = 22°$, $c = 0.044$ MPa, $E_s = 33$ to 40 MPa; the full depth of this stratum was not determined.

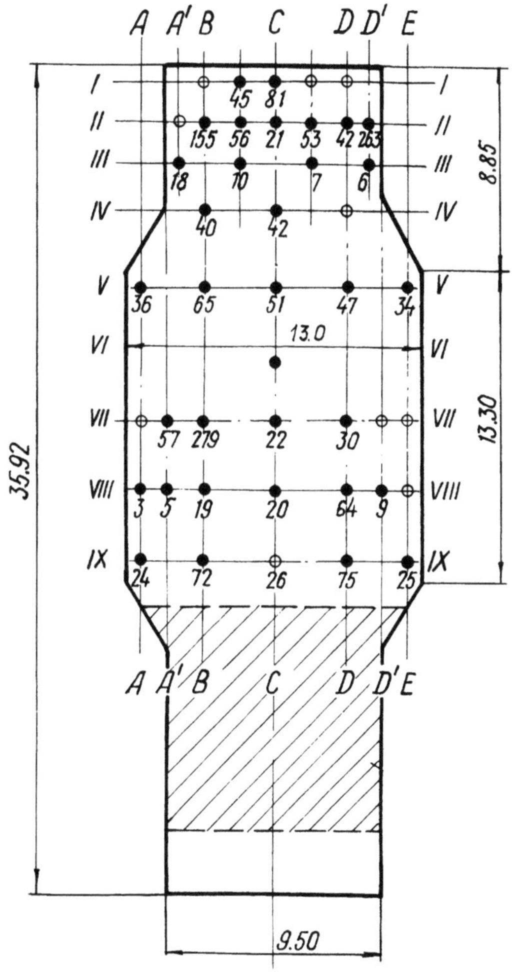

Figure 8-14. Thermopower generating station at Kostroma, Russia. Outline of turbine/generator foundation slab and layout of contact soil pressure measuring cells. *Note*: Shaded area is not instrumented.

The foundation bedding included a 300-mm-thick layer of crushed stone overlain by a 300-mm-thick layer of medium sand. Similar to the previous case histories, SDKS-type cells, preinstalled into concrete blocks (Figure 8-6), were placed on thoroughly leveled sand at predetermined locations, after which a 20-cm-thick layer of a lean concrete was carefully installed. This was followed by the installation of structural slab reinforcement and casting of the mass concrete. A total of 44 instruments were installed in nine transverse and seven longitudinal sections of the slab. This is illustrated in Figures 8-14 through 8-16.

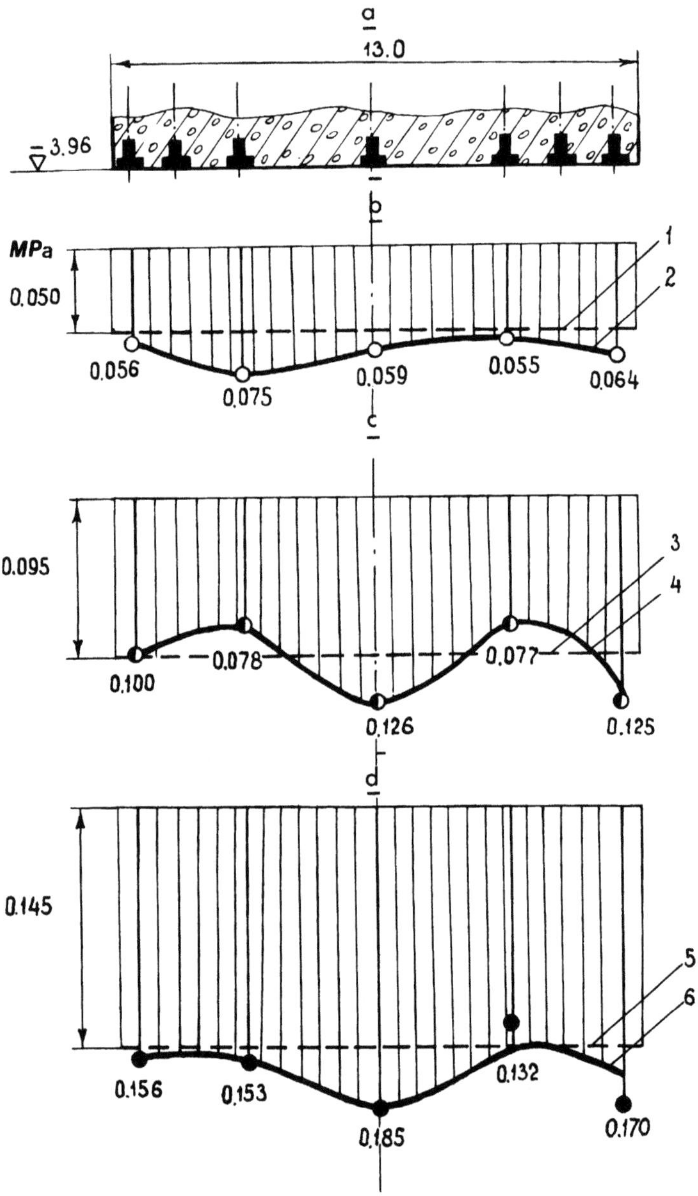

Figure 8-15. Thermopower generating station at Kostroma, Russia. Averaged contact soil pressure diagrams at cross sections V-V through IX-IX (from Figure 8-14). (a) Instrument distribution along slab cross section; (b) soil reaction from slab weight; (c) soil reaction from weight of superstructure; (d) soil reaction from total weight. 1, 3, and 5. Uniformly distributed (average) load due to structural weight(s); 2, 4, and 6. actually measured soil pressures due to respected loads.

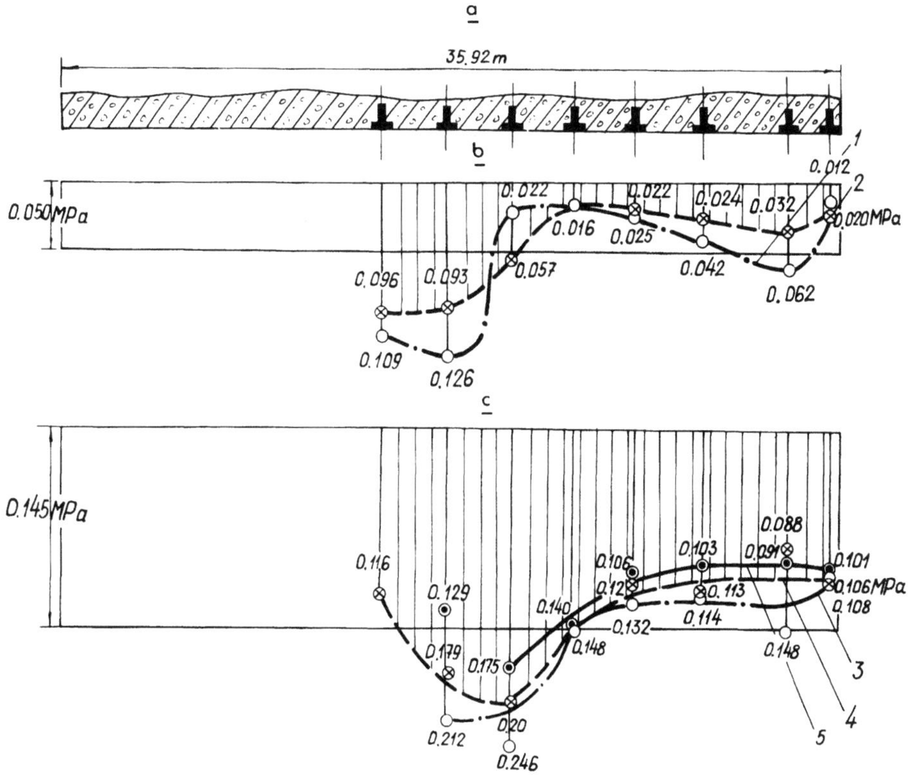

Figure 8-16. Thermopower generating station at Kostroma, Russia. Averaged contact soil pressure diagrams in longitudinal sections B-B through D-D (from Figure 8-14). (a) Instrument distribution along longitudinal section; (b) soil reaction diagrams attributed only to weight of slab; (c) soil reaction diagrams attributed to total weight of foundation structure. 1 and 3. Measured soil pressures at cross section C-C; 2 and 4. average readings at cross sections B-B, C-C, and D-D; 5. soil pressure readings 34 and 41 months following start of turbine/generator operation.

The "flexibility index" of the slab in the transverse direction, t, was calculated from $t \simeq 10(E_s b^3/E_f h^3)$ (Gorbunov-Possadov and Malikova 1973) and found to be less than 1. This result indicates that this slab is very stiff. Here E_s and E_f are, respectively, the moduli of deformation of the soil and the foundation structure, and b and h are, respectively, the width of the slab and its thickness. By the same criteria, in the longitudinal direction the slab's "flexibility index" is much greater than unity, i.e., $t >> 1$. This defines this slab as nonstiff in the longitudinal direction. The measured soil reaction distributions for the different construction stages are shown in Figures 8-15 and 8-16.

The averaged contact soil pressure at sections V-V through IX-IX (see Figure 8-14) are given in Figure 8-15. The soil pressure distributions in the longitudinal direction (sections B-B through D-D) are shown in Figure 8-16. The results of this contact soil pressure monitoring program indicate that the qualitatively obtained soil pressure diagrams correspond to those obtained by finite element analysis based on realistically determined soil deformation parameters data.

The results of all investigations described in this chapter also indicate that "theoretically" very stiff structures are still flexible enough to affect their interaction with foundation soil.

9

Stiff Foundations on Cohesive and Nonhomogeneous Soils

This chapter contains several characteristic examples that illustrate the interaction between stiff foundations of miscellaneous construction with soils of cohesive and nonhomogeneous nature.

9.1 Perforated Stiff Foundation System on Cohesive Soils

This economical type of foundation, which is designed to carry heavy point loads, e.g., from a column, is comprised of two perforated reinforced concrete blocks 2.25 × 0.75 m placed side by side and overlain by a stiff pedestal made of a reinforced concrete block 1.5 × 0.75 × 0.35 m (Figure 9-1). In this figure the pedestal is outlined by the dashed line.

The foundation soil is represented by a loam having a density of 18.1 kN/m^3 and a water content of approximately 11%. A year prior to the tests that have been carried out on foundation in question, the foundation soil was loaded by a static load in increments of 0.1, 0.2, and 0.3 MPa. This was required to study the stress distribution in the mass of soil. Subsequently, in the process of extracting the pressure cells that were used in this investigation, the upper portion of this foundation soil was loosened to a depth of approximately 40 cm.

In this work flat membrane pressure cells were installed 150 mm from the underside of the foundation block and overlain by sand and a compacted crushed stone bedding as shown in Figure 9-1. The arrangement of these cells in plan is also shown in this figure. The foundation system was loaded by means of a hydraulic jack placed against a reinforced concrete test beam.

The load was applicd in increments of 0.05, 0.1, 0.2, and 0.3 MPa. In conjunction with the stresses in the soil, settlement of the system and stresses in the block reinforcing steel was measured. This was done in Sections I-I through III-

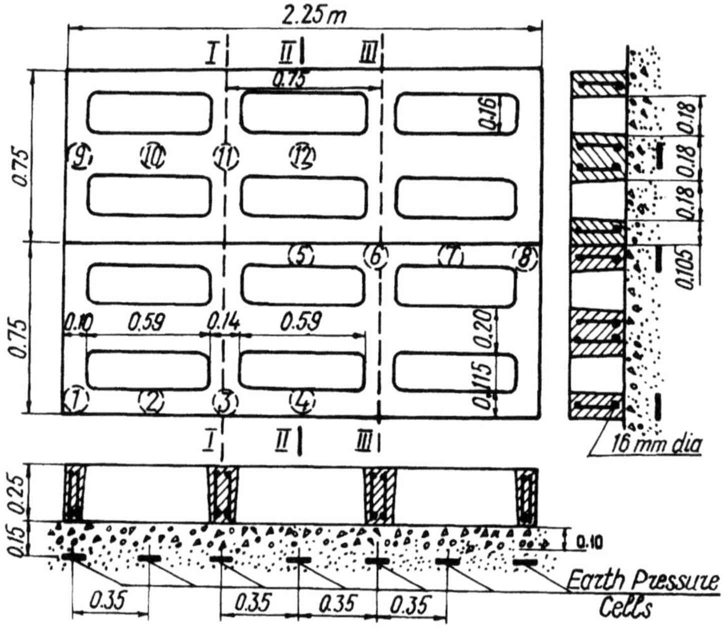

Figure 9-1. Perforated concrete foundation system on cohesive foundation soil; plan and typical cross sections. *Note*: Position of concrete pedestal 1.5 × 0.75 m is outlined by the dashed line.

III (Figure 9-1), where the bending moment was expected to have the greatest magnitude.

Measured values of soil pressure at different phases of load application are shown in Figure 9-2. Also shown in this figure by the interrupted lines are the corresponding averaged applied loads (in MPa). The calculations, based on measured deflections of the blocks and stresses in their steel reinforcement, demonstrate good correlation with obtained soil reaction diagrams at all loading stages. From the results of these tests one may conclude that the use of just uniformly distributed loads for foundation design (which is the case in most practical designs) may result in an underdesigned structure due to stress concentrations at the foundation edges. This consequently results in larger values of bending moments in the foundation blocks.

9.2 Solid Concrete Slab Constructed Upon a Nonhomogeneous Foundation Soil

A 1.0-m-thick concrete slab was constructed in Kiev, Ukraine, to support loads from a 12-storey apartment building (Figure 9-3). A typical cross section and

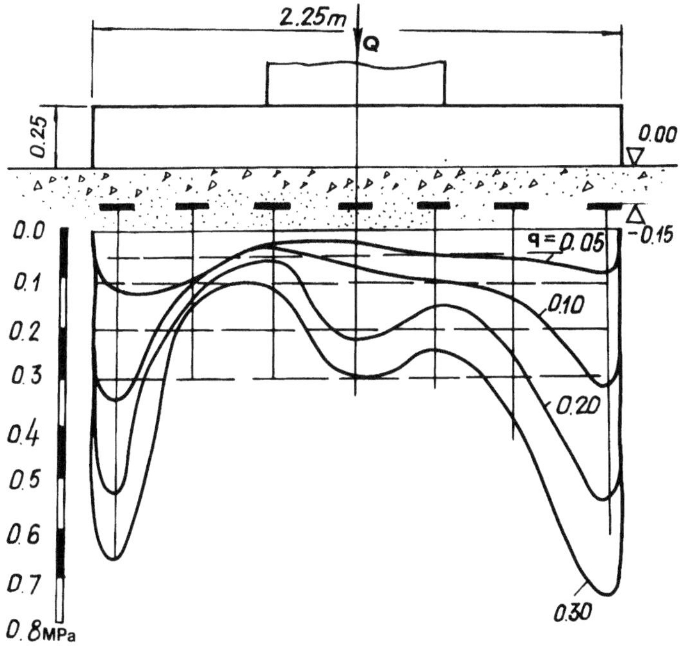

Figure 9-2. Perforated concrete foundation system on cohesive soil. Soil reaction diagrams to equivalent load, q, equal to 0.05, 0.1, 0.2, and 0.3 MPa.

plan of the basement section of this building are shown in Figure 9-4. The foundation soil is composed of miscellaneous deposits that were dumped there a century ago to fill a relatively narrow, but very deep, ravine that existed at the construction site. The material deposited was dumped on top of a vegetative humus ground. The fill within the ravine consists from the top down of the following:

- a low-water-content sand and sandy loam with inclusion of construction debris; thickness varies from 0.6 to 5.3 m
- a mixture of loam, sandy loam, and sands of miscellaneous compositions; thickness varies from 1.2 to 12.2 m
- a mixture of loam and humus; thickness varies from 0.2 to 1.6 m
- a loamy loess; thickness is equal to approximately 8 m
- loams of miscellaneous compositions and sandy loams; thickness varies from 0.5 to 5 m.

This fill was underlain by the original soil as previously noted. The deformation modulus of each soil was determined in situ. It was found that the value of the average deformation modulus of the deposited soils was equal to approximately 15 MPa and that of the original soil was equal to approximately 25 MPa The

Figure 9-3. Apartment building in Kiev, Ukraine. General view.

groundwater table was located at approximately 14 m from grade. Investigation of the soil in close proximity to the building's cross section, in which the contact soil pressures were monitored, indicated that at this location the middle portion of the foundation slab was founded upon deposited soils, but at the slab's edges, the distance from its underside to the original soil, which is represented basically by sandy loams, was only 0.35 to 0.5 m. A detailed account of the soil conditions at this particular project is given in Lazebnik (1973).

The contact soil pressures were measured by an SDKS-type pressure cell, as described in Chapter 4. The soil reaction diagrams measured at different stages during the building construction are shown in Figure 9-4. It must be noted that because the soil foundation load did not increase uniformly in the longitudinal direction (due to construction schedules the load was constantly moving from one end of the building to the other), the soil pressure readings were affected by the corresponding slab deformations. The soil pressure readings indicated in the soil reaction diagrams of Figure 9-4 represent maximum values of the soil pressure attributed to the action of the maximum load at the cross section where pressure cells were installed.

Soil pressure monitoring in this case history indicates the fact that for a long

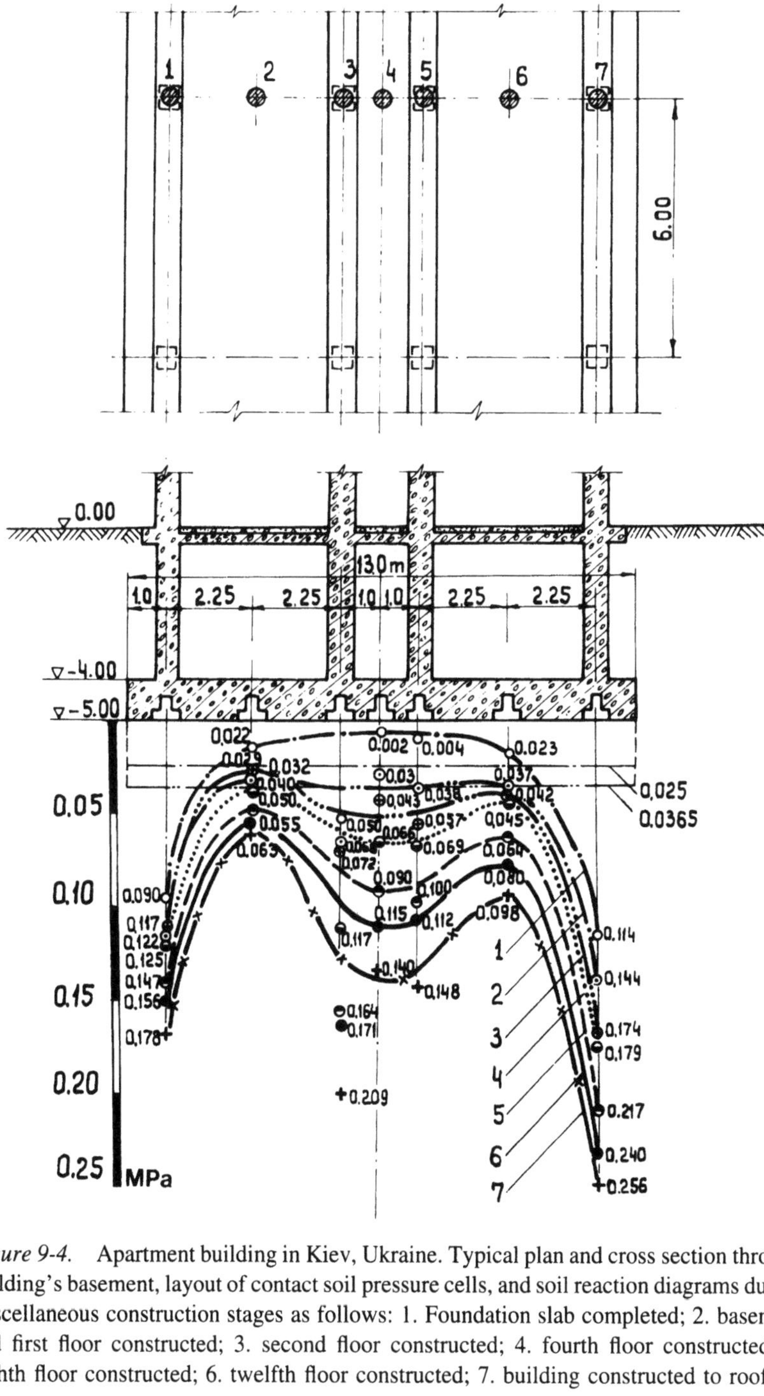

Figure 9-4. Apartment building in Kiev, Ukraine. Typical plan and cross section through building's basement, layout of contact soil pressure cells, and soil reaction diagrams during miscellaneous construction stages as follows: 1. Foundation slab completed; 2. basement and first floor constructed; 3. second floor constructed; 4. fourth floor constructed; 5. eighth floor constructed; 6. twelfth floor constructed; 7. building constructed to rooftop.

and rather flexible (in the longitudinal direction) footing the foundation soil is exposed to periodic load fluctuations. This is the result of nonuniform load application during the construction process. In general, the measured soil reaction diagrams conform to those obtained by theory applied to slabs performing on elastic foundations. Similar results can be obtained by use of the finite element method of analysis, provided that boundary conditions are properly identified and soil properties are correctly determined.

It should be noted that stress values obtained at the end of the building construction were periodically monitored during a 12-year period subsequent to building commissioning. During this period of time no substantial changes in instrument readings were detected.

9.3 Soil Pressure from a Water Reservoir

This structure is shown in Figure 9-5. It is constructed from prefabricated concrete components and has a capacity of 2000 m^3. The plan dimensions of this structure are depicted in Figure 9-6. The reservoir is filled with fresh water to a depth of 5 m. Hence, it produces a maximum uniform pressure of 0.05 MPa that is transmitted to the founding soil through a relatively thin (120-mm-thick) cast in situ reinforced concrete slab.

The foundation soil is comprised of loam with a water content (w) of 4.5% and a plasticity index (I_p) of 17%. The granulometric composition of this soil is

Figure 9-5. General view of concrete water storage reservoir.

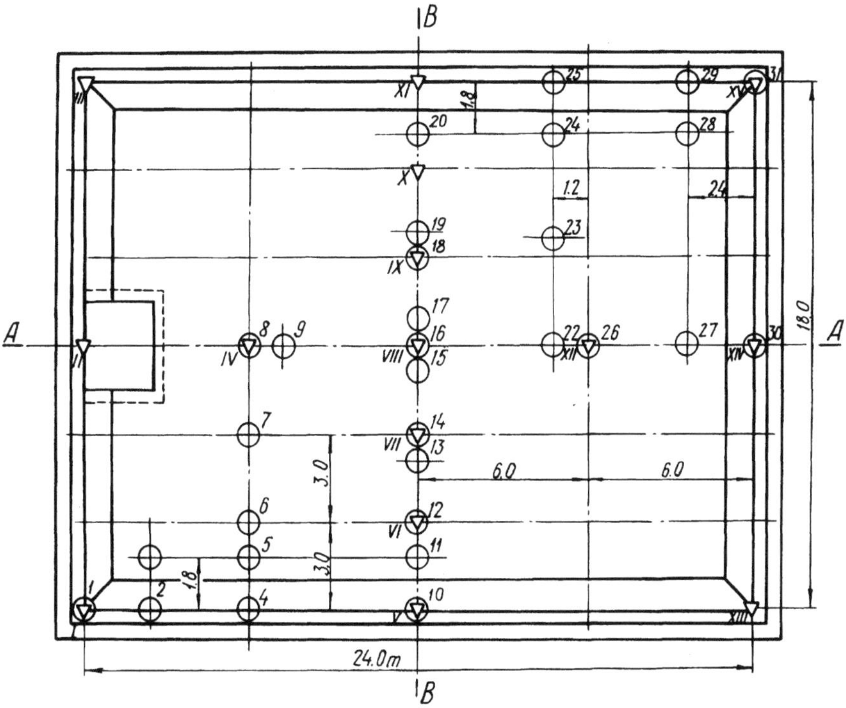

Figure 9-6. Concrete water storage reservoir. Layout of soil pressure cells and survey marks. ○—pressure cell; ▽—survey mark.

given in Table 9-1. The soil reaction was measured by vibrating wire type pressure cells at 30 different locations. The layout of these cells is shown in Figure 9-6. Measurements of reservoir cover settlements were made at 15 locations, as indicated in Figure 9-6. The pressure cells were installed on the surface of the foundation soil through openings made in a previously installed layer of lean concrete; subsequently the foundation slab was cast upon this lean concrete.

The reservoir was filled with water and emptied twice, and both times contact soil pressures and settlements resulting from the weight of the water only were measured. The water was introduced in three stages, each equal to one-third of the total full volume. Increases in cell readings at all three stages were close to linear.

Table 9-1. Granulometric Composition of Foundation Soil Underlying Structure of Water Reservoir

Particle diameter (mm)	0.5–0.25	0.25–0.05	0.05–0.01	0.01–0.005	< 0.005
Particle content (% weight)	0.6	5.2	46.9	12.3	35.0

The measured soil reaction diagrams are depicted in Figure 9-7. From this figure it is seen that the soil pressure diagrams measured during the first and second fillings are almost identical. The convex shape of these diagrams may be explained by the fact that the central part of the foundation soil was densified by heavy lift equipment which was used for erecting prefabricated reservoir walls. Otherwise, bearing in mind the flexibility of the slab, the soil reaction diagram should be rectangular. This is confirmed by an investigation conducted by Lazebnik et al. (1973) on a similar structure constructed on a noncompact cohesive soil.

9.4 Influence of Nonuniform Densification on Shape of Soil Reaction Diagram

The nonuniform densification of foundation soils was investigated by using stiff round and rectangular large-scale models placed on foundation soils composed of medium-grain sand. The shape of the soil reaction diagram significantly affects the design of the structural foundation. In practical cases of foundation design, the shape of this diagram affects the values of bending moment and shear force that are used to determine the size and properties of structural elements included within the foundation structure.

From previous discussions it is clear that the state of the soil/structure interaction is a function of both the stiffness of the structure, e.g., its deformation characteristics, and the density of the founding soil. Hence, by changing the properties of either the structure or the soil, it may be possible to control the distribution of contact soil pressures at the soil/structure interface. This idea has been around for a long time (Kravtsov and Evdokimov 1957; Evdokimov 1962; Borodachev 1972; Sorotchan et al. 1981).

In order to control the shape of the soil pressure diagram the above researchers

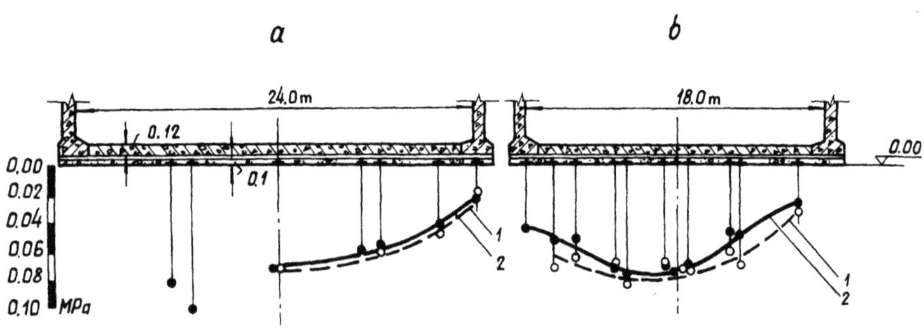

Figure 9-7. Concrete water storage reservoir. Soil reaction diagrams at A-A and B-B (see Figure 9-6). (a) Longitudinal cross section; (b) transverse cross section. 1. First loading cycle; 2. second loading cycle.

basically proposed either to make changes in the founding soil, e.g., introduce greater density in the middle part of the foundation and lesser density at the edges of the structure and/or to change the shape of the interface surface, e.g., make it convex rather than flat.

To review some of these recommendations, this writer and co-workers carried out a series of large-scale experiments using models of different configurations as noted earlier in this section.

9.4.1 Stiff Circular Concrete Block 2.2 m in Diameter

A general view and typical cross section of this block and the layout of the contact soil pressure cells are illustrated in Figures 7-4 and 9-8. The soil pressure cells in this block were installed radially at equal distances from the center of the block (e.g., 0.28, 0.56, 0.8, and 1.01 m). The block was tested on an unevenly densified foundation soil. The central part of the test block (approximately one-third of its underside area) was placed on a sand foundation having a density of 17.1 kN/m^3 (D_r = 66%). Its perimeter portion was placed on less dense sand having a density of 16.5 kN/m^3 (D_r = 45%). Deep soil zones (0.5 and 1.0 m) had a greater density.

The test load was applied in increments of 0.05, 0.1, 0.2, and 0.25 MPa, and

Figure 9-8. Round concrete block 2.2 m in diameter. Installation of soil pressure cells.

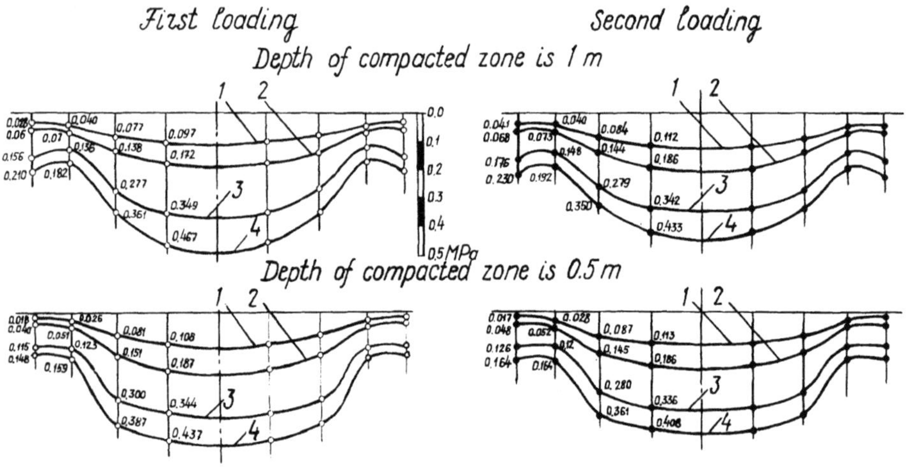

Figure 9-9. Round concrete block 2.2 m in diameter placed on nonuniformly dense foundation soil. Soil reaction (1 through 4) diagrams corresponding to outside pressure equal to, respectively, 0.05, 0.1, 0.2, and 0.25 MPa.

the block was loaded in two cycles. The resulting soil contact pressure diagrams are shown in Figure 9-9. As is seen from this figure, the shapes of these diagrams differ substantially from those presented in Chapter 7 (Figure 7-5). Those from Chapter 7 were obtained from tests on the same circular block, but placed on a uniformly dense foundation soil.

It is seen that in the case where the foundation soil is denser at the middle part of the footing, the shape of the soil reaction diagram has a distinctly convex shape (Figure 9-9). However, in the case of a uniformly dense foundation soil, the soil reaction diagrams have a wave-like shape (Figure 7-5).

From Figure 9-9 it follows that an increase in depth of the soil compacted zone results in an increased soil pressure in the middle part of the footing and a corresponding decrease at its edges. It is also obvious that under the described test conditions soil pressure ordinated related to the second loading cycle are slightly greater at the footing edges and correspondingly smaller in its middle portion.

9.4.2 *Stiff Rectangular Concrete Block 3 × 1 × 0.6 m*

The plan, typical cross section of this block, and layout of the contact soil pressure cells installed at the block's underside are shown in Figures 7-8 and 7-9. In these experiments this block was tested under the following two soil conditions:

1. The middle section of the block (1 m long) rests upon a dense foundation soil, and its ends (each 1 m long) sit upon less dense soil.

2. The middle section of the block (2 m long) rests upon a less dense soil, and its ends (each 0.5 m long) sit upon a soil of greater density.

The soil densities (e.g., less dense and more dense) in these tests were similar to those described in previous experiments.

In the first series of tests the block was loaded by a load equal to 0.5, 0.1, 0.2, and 0.3 MPa, and in the second series the corresponding loads were equal to 0.1, 0.2, and 0.3 MPa. Similar to the previous series of tests with round blocks, the rectangular block was loaded in two cycles, i.e., up to the maximum load in the above increments, then unloaded and loaded again up to the same maximum load. Also, similar to the previous tests, the foundation soil was densified to a depth of 0.5 and 1.0 m. The results of these tests are shown in Figures 9-10 and 9-11.

As is seen from these figures, and similar to previous tests, the denser areas of the founding soil attract greater values of soil pressure. Under identical load conditions, the soil reaction diagram shape and the value of the stress ordinate in both cases are distinctly different, and in both cases the shapes of the soil reaction diagrams are substantially different from those obtained from load tests on the same block placed on soils of uniform density (Figures 7-10 and 7-11).

From the experimental results described in this section and comparison with the previous results obtained from identical concrete blocks, but placed on uniformly densified soils, the indications are that by creating a hardened or densified soil

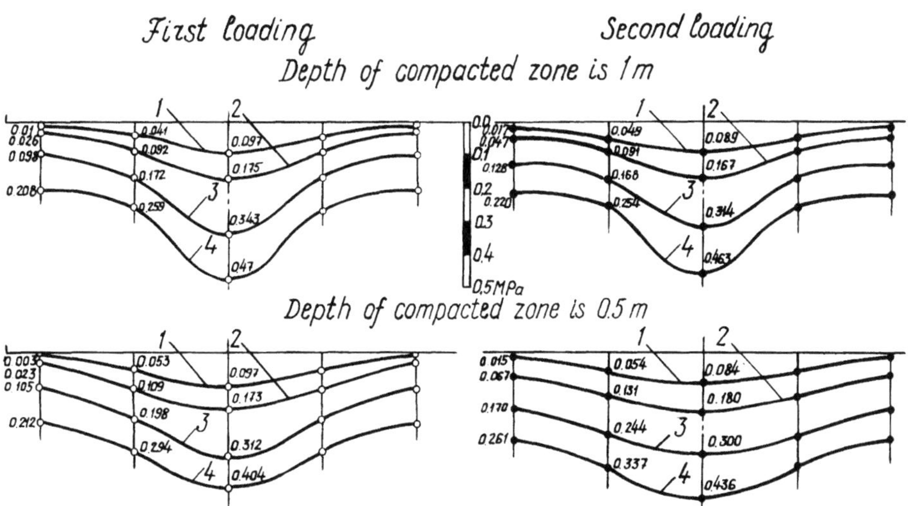

Figure 9-10. Rectangular concrete block 3 × 1 × 0.6 m placed on nonuniformly dense foundation soil with denser soil area located at block's middle section. Averaged soil reaction diagrams along block's longitudinal axis: 1 through 4. diagrams corresponding to outside pressure equal to, respectively, 0.05, 0.1, 0.2, and 0.3 MPa.

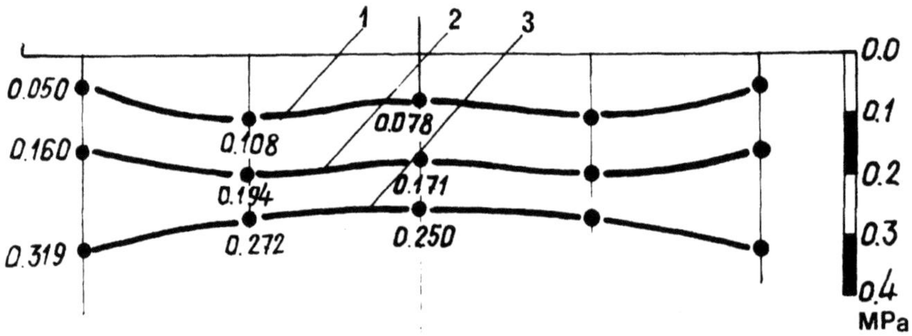

Figure 9-11. Rectangular concrete block 3 × 1 × 0.6 m placed on nonuniformly dense foundation soil with denser soil areas located at block's ends. Averaged soil reaction diagrams along block's longitudinal axis: 1 through 3. diagrams corresponding to outside pressure equal to, respectively, 0.1, 0.2, and 0.3 MPa.

core at designated locations, the engineer can effectively control the shape of the soil reaction diagram and, by doing so, may affect the design of the foundation structure. The results of the preceding tests also suggest that cyclic loads have practically no effect on the shape of the soil reaction diagram.

Finally, it should be noted that artificially created dense zone(s) can, to some extent, affect settlement of the footing. In the previously discussed cases, the footing settlement during the first cycle of loading was 9 to 11% greater in comparison to a similar footing placed on a uniformly densified soil. In the second loading cycle this effect decreased to approximately 4%.

10

Lateral Earth Pressure "At Rest"

10.1 Introduction

Under conditions where there is no lateral strain within the ground mass, the value of the lateral soil pressure is commonly called the *lateral earth pressure at rest* (K_0). Sometimes it is also defined as the *neutral lateral earth pressure* or the *lateral earth pressure at consolidated equilibrium.* The ratio of lateral to vertical earth pressure in this "no lateral strain" condition is termed the *coefficient of earth pressure at rest,* $K_0 = \sigma_3/\sigma_1$. Here σ_3 is the principal horizontal stress and σ_1 is the principal vertical stress, both in a soil mass having a cylindrical form of test apparatus.

According to the theory of elasticity, the coefficient of earth pressure at rest depends solely on the value of Poisson's ratio ν, i.e., $K_0 = \nu/(1 - \nu)$ (Tschebotarioff 1973). In some practical situations, e.g., the design of rigid unyielding walls such as gravity-type walls constructed on stiff foundation soils or basement walls backfilled with sand, K_0 will typically be equal to 0.4–0.5. In clays of normal sensitivity, K_0 is usually 0.5–0.6.

In granular material the magnitude of K_0 depends on the amount of frictional resistance mobilized at contact points between particles (Lambe and Whitman 1969). When a granular soil is loaded for the first time, the frictional forces at the contacts act in such a direction that σ_3 is less than σ_1, i.e., $K_0 < 1$. Jaky (1944), on the basis of his experiments, suggests that in granular soils the coefficient of lateral earth pressure at rest is best represented by $K_0 = 1 - \sin\phi$, where ϕ is the angle of internal friction.

By running a series of odometer tests in which not only the vertical effective stress but also the lateral effective stress was measured, Brooker and Ireland (1965) developed relationships for K_0 as an average for all normally consolidated clays: $K_0 = 0.95 - \sin\bar{\phi}$, where $\bar{\phi}$ is the effective angle of internal friction. For normally consolidated clays, Alpan (1967) recommends the empirical formula $K_0 = 0.19 + 0.233 \log PI.$, where $PI.$ is the plasticity index of clay, in percent. Since the early 1920, numerous experiments have been conducted by many workers in order to determine a credible value of K_0. Terzaghi (1920) found that K_0 does not depend on whether the soil is loose or dense. For sands he found $K_0 = 0.42$ and for clays $K_0 = 0.7$ to 0.75. Fedorov and Malyshev (1954) found that K_0 in loose soil has a greater value than in dense soils. Rowe (1954) found that the value of K_0 decreases with increases in vertical pressure σ_1. It should be noted that the investigators who tested soils in triaxial cells usually obtained larger values of K_0 than those who tested loose soils. For more information on K_0 values, the interested reader is referred to Clayton et al. (1993).

This author, with co-workers, undertook a large-scale model investigation to determine the value of K_0 in granular soils using the calibration chambers, soil pressure measuring systems described in the preceding chapters and some special proprietary devices.

10.2 Experimental Determination of K_0 in a Granular Soil Mass

Clean, dry quartz sands of various granulometric compositions were used in these experiments: two types of coarse granular sands, two kinds of fine granular sands, and a medium-sized granular sand. The principal stresses in the soil mass were determined using stiff flat pressure cells. Under test conditions it is possible to assume that $\sigma_1 = \sigma_z$ and $\sigma_2 = \sigma_3 = \sigma_r$. The first series of tests was carried out in order to test the basic criteria of the experiment methodology.

At this early stage the medium sand was tested in the smaller calibration device, the SCC (for details consult Chapter 3), by means of nine small-diameter ($d = 26$ mm), stiff cells. The membrane deflection of these cells did not exceed 1/2000 d at maximum pressure. In these tests the value of K_0 was determined from two experiments as the σ_3/σ_1 ratio as the average of 18 measurements for dense ($D_r > 67\%$) and loose ($D_r < 30\%$) sands. In order to obtain a value of σ_1 in a mass of soil, the cells were arranged horizontally and loaded to pressures of 0.5 and 0.1 MPa. Then, in order to measure a σ_3 value, the same cells were rearranged vertically and loaded with similar loads. In all tests the soil's density was carefully controlled to ensure comparable soil conditions in all relevant tests.

In order to account for effects of possible displacement of pressure cells in the mass of soil during load application, the cells were installed in three different configurations: "cross," "circle" with pressure-bearing membranes facing the wall of the chamber, and "circle" with pressure-bearing membranes facing the chamber's center. From these tests the value of K_0 was found to vary from 0.53 to 0.66 with larger values related to loose sand (Lazebnik 1967). It must be noted that in this preliminary series of tests the results were quite scattered. This apparently was attributed to effects of the small diameter and high stiffness of the soil pressure measuring instruments used. Subsequently, these tests were repeated, this time measuring the soil pressures with larger, stiffer cells of the PDGS type with hydraulic multiplicators (for details on these cells, consult Chapter 1).

To measure $\sigma_3 = \sigma_r$, the cells were arranged in the calibration chamber as depicted in Figures 10-1 and 10-2. The depth of sand placed in the calibration chamber, H_s, was equal to 260 mm. All tests were repeated 4 to 5 times to obtain reliable data. The use of large-diameter instruments as well as the duplication of cell loadings under identical soil conditions resulted in significantly reduced scatter in the results. Characteristic results of these experiments are shown in Figure 10-3.

In these tests, similar to the previous experiments, the value of K_0 decreased with an increase in σ_1 in both loose and dense, medium- and coarse-grained sands, and the value of K_0 was greater in a dense sand. It should be pointed out that during the second load cycle the values of K_0 slightly decreased. As is seen from Figure 10-3, K_0 values in this set of tests were remarkably smaller than those obtained in the preliminary series of experiments where K_0 was equal to 0.53–0.66. This suggests that cell readings can be influenced by cell diameter relative to the size of the calibration chamber, as well as by the depth of soil in the calibration device.

This phenomenon was investigated in the next series of tests. These were carried out using PDGS-type cells having $d = 120$ mm tested in the large calibration chamber (LCC). The cells were installed in a pattern similar to that used in previous experiments. Additionally, tests with only one cell installed at the center of the calibration chamber was conducted. The depth of sand, H_s, in the test chamber was equal to 260 mm, as in previous tests, and additionally all tests were repeated at $H_s = 400$ mm. Because the diameter of the LCC is more than

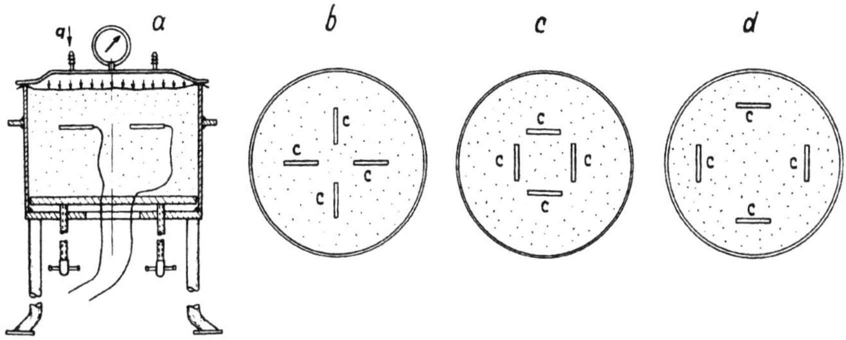

Figure 10-1. Investigation of K_0 value. (a) Typical cross section through calibration chamber with pressure cell installed horizontally to measure vertical principal stress σ_1; (b), (c), and (d) arrangements of vertically installed cells to measure horizontal principal stress σ_3; c (designates orientation of cell pressure-bearing membrane.

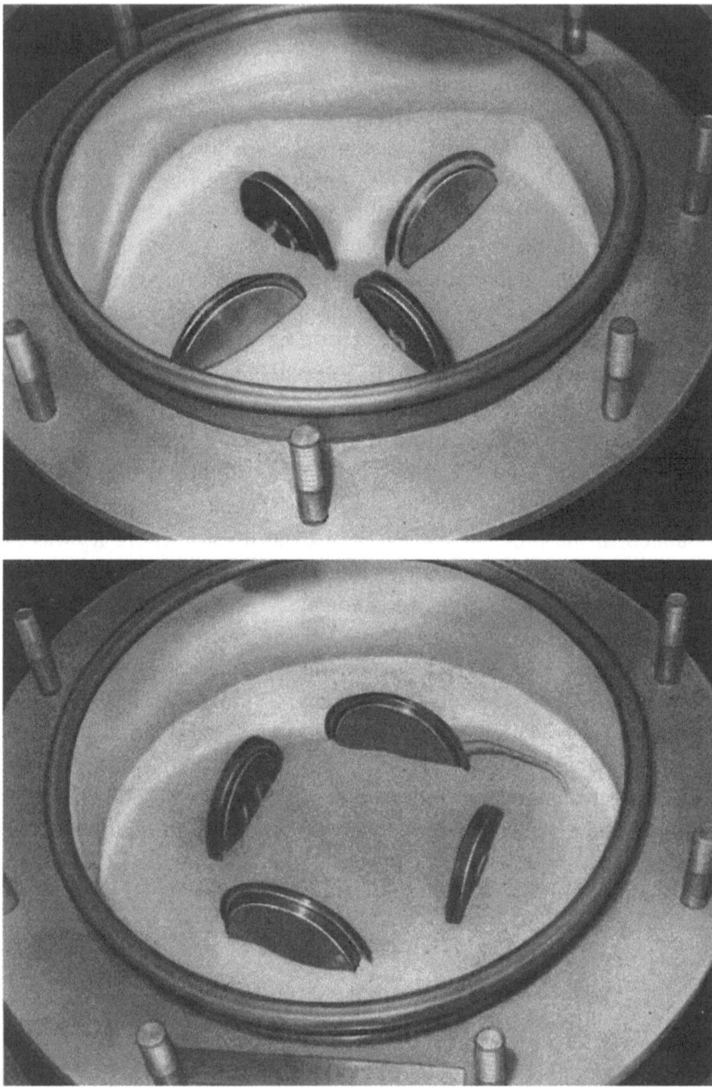

Figure 10-2. General view of calibration chamber with pressure cells installed in different arrangements.

twice the diameter of the SCC, under the H_s = 260 mm condition cell readings are comparable to those obtained in the first set of tests.

The test results are shown in Figures 10-4 and 10-5. These figures show the values of K_0 to be in the range of 0.35 to 0.535, much larger than those obtained in experiments carried out in the SCC using identical instruments. This confirms

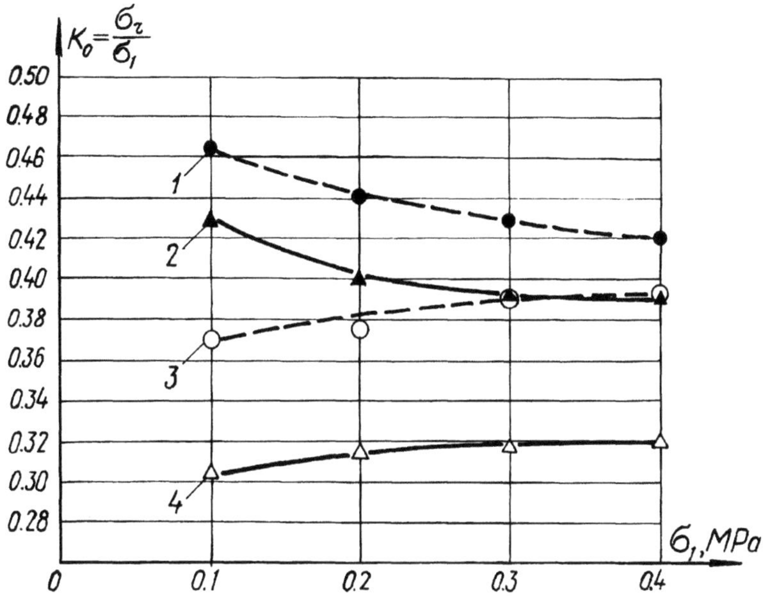

Figure 10-3. K_0 value as function of vertical pressure (σ_1) and soil granulometry and density. 1 and 3. In medium dense and loose sand, respectively; 2 and 4. in coarse dense and loose sand, respectively.

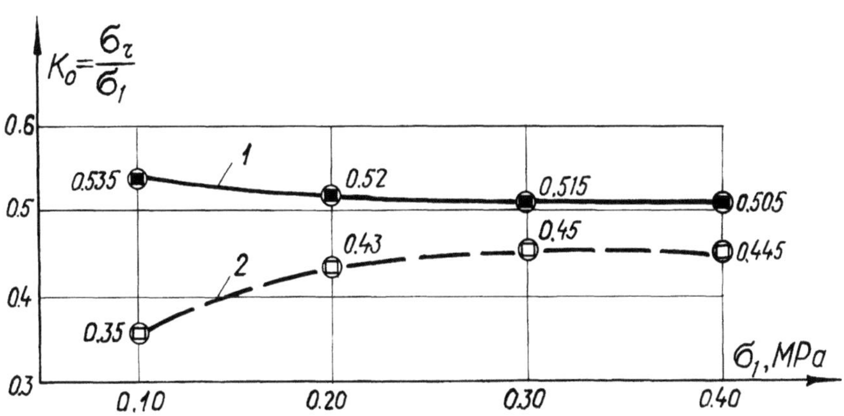

Figure 10-4. Results of experiments on determining K_0 carried out in large calibration chamber LCC (for details consult Chapter 3). Cells are arranged according to scheme depicted in Figure 10-1c; depth of sand, H_s = 400 mm. 1. In dense medium sand. 2. in loose medium sand.

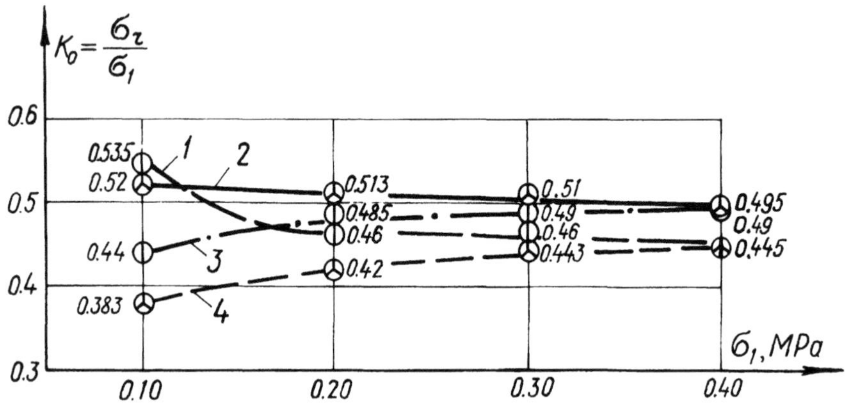

Figure 10-5. Results of experiments on determining K_0 carried out in large calibration chamber LCC (see Chapter 3). Conditions are similar to those defined in Figure 10-4 with addition of tests with a single pressure cell. 1 and 3. Single pressure cell in dense and loose soil, respectively; 2 and 4. cells arranged according to scheme depicted in Figure 10-1b, in dense and loose soil, respectively.

that the soil pressure measured in a mass of soil that is confined within the calibration chamber is influenced by the cell-to-chamber diameter ratio. Some characteristic data on K_0 values for loose sands as a function of soil depth (H_s) within the large calibration chamber LCC is given in Table 10-1. From this data it can be seen that at $H_s = 400$ mm, the K_0 values are smaller than those measured at $H_s = 260$ mm. This is most likely a manifestation of the increase in friction value between the soil and internal face of the chamber, which may decrease the actual soil pressure on the cell(s).

The data obtained from these and other miscellaneous related experiments makes it possible to correct the $K_0 = f(\sigma_1)$ curves depicted in Figures 10-4 and 10-5. A characteristic example of corrected curves $K_0 = f(\sigma_1)$ is illustrated in Figure 10-6. It is seen that irrespective of orientation of the cell's pressure-bearing membrane, the recorded values of K_0 are very close to each other for each respective soil density and stress σ_1. Besides other things, this indicates that the horizontal displacement of the cell within the soil mass in the test chamber is either nonexistent or negligible. From Figure 10-6 it follows that K_0 for medium sand with increasing σ_1 tends to acquire the constant value of 0.5: increase in σ_1 in coarse sand would result in $K_0 = 0.4$ to 0.43.

10.3 Coefficient of Lateral Soil Pressure, *K*, at Interfaces with Stiff Walls

An investigation into this matter was conducted with medium sand placed in rings of different diameters and depths (Figures 5-3 and 5-4). In this investigation

Table 10-1. Relationship Between K_0 *and Height of Loose Medium Sand* H_s *in Calibration Chamber LCC*

Pressure σ_1 (MPa)	Values of Coefficient K_0 at Soil Height H_s (mm)				In Percents, Taking Value of K_0 at H_s = 260 mm as 100%		
	260		400				
	I Cycle	II Cycle	I Cycle	II Cycle	I Cycle	II Cycle	Mean Value
0.1	0.466	0.36	0.40	0.32	86	89	87.5
0.2	0.51	0.386	0.426	0.37	84	96	90.0
0.3	0.49	0.43	0.416	0.38	85	88	86.5
0.4	0.48	0.44	0.456	0.40	95	91	93.0

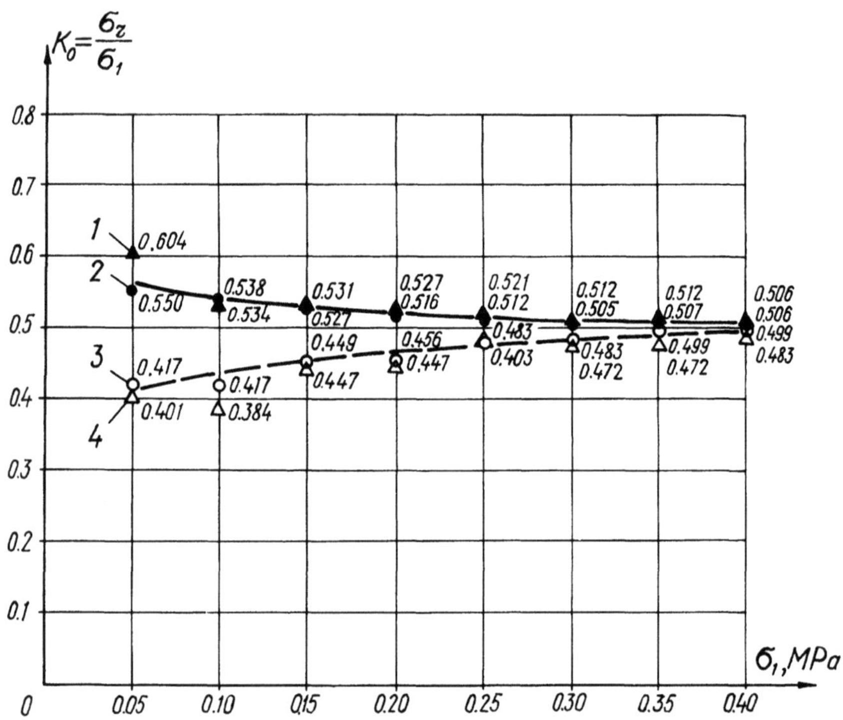

Figure 10-6. Corrected relationship $K_0 = f(\sigma_1)$ for medium sand. Tests are carried out in LCC and are arranged according to Figure 10-1c. 1 and 2. In dense sand; 3 and 4. in loose sand. ○—pressure-bearing membrane faces LCC wall; △—membrane faces center of LCC.

values of $K = \sigma_r/\sigma_z$ obtained in rings 300 and 260 mm in diameter were in close agreement. Here σ_r is the horizontal principal stress and σ_z the vertical stress. The results of these tests are shown in Figure 10-7. This figure shows that the value of K decreases with increases in σ_z irrespective of soil density. This may suggest that the ring's stiffness is insufficient, and its radial expansion, although very small, does have some effect upon the measured K value.

In order to exclude any doubt in the outcome of these tests, a special, very stiff, thick-walled, square device of composite (steel/concrete) construction with polished inner surfaces was used to obtain the K value. The lateral soil pressure in this device was measured by four SDKS-type pressure cells installed in the walls of the device. Pressure was applied to the soil placed inside the device chamber by a hydraulic press via a very stiff steel cover plate. In the former case, settlement of the cover plate was measured by means of standard dial gages. Two modifications to the special square device were used to further investigate K values. Both had chambers 150 mm deep, however, one chamber was square 310 × 300 mm and the other was round with a diameter of 470 mm. Typical

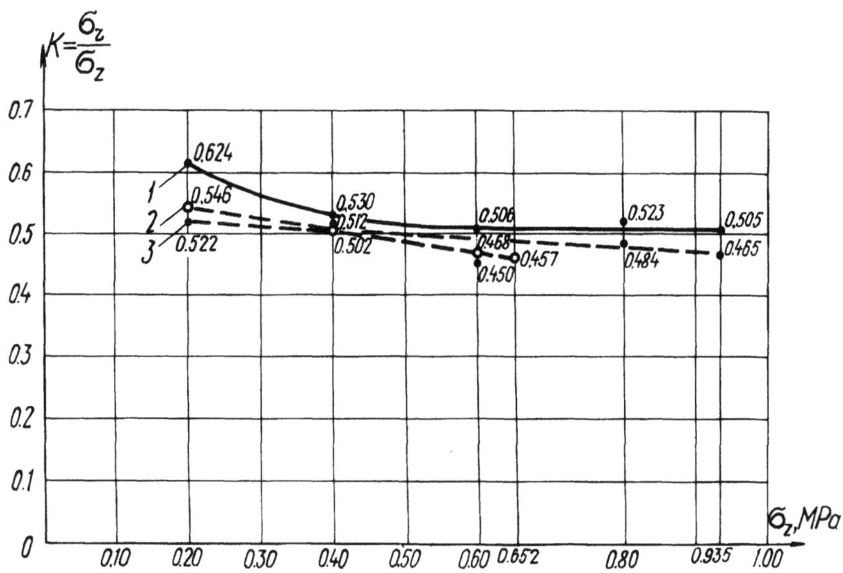

Figure 10-7. Values of $K = \sigma_r/\sigma_z$ obtained in large rings. 1 and 3. Ring 260 mm dia., compacted and loose sand, respectively; 2. ring 300 mm dia., loose sand.

cross sections through these devices and their general views are shown in Figures 10-8 through 10-10. Vertical loads on the carefully placed and leveled soil were applied in steps of 0.05 and 0.1 MPa. The value of lateral pressure was determined as an average from readings from four contact pressure cells. Figure 10-11 shows results of $K = f(\sigma_z)$ for a medium quartz river sand and for fine crushed stone obtained from tests in the square device depicted in Figure 10-8a.

The granulometric composition of the crushed stone and some of its other geotechnical parameters are given in Table 10-2. Both the medium sand and the fine stone were tested in a loose ($D_r < 33\%$) and a dense ($D_r > 0.67\%$) state. As is seen from Figure 10-11, for both materials tested with increasing σ_z, the difference in K values between loose and dense soils decreases, and at $\sigma_1 = 0.6$ MPa it becomes negligible. It is also seen that K values for fine-grained soils are larger than those for coarse-grained material.

In order to review the effects of load application on K values obtained in a square device, the tests described above were repeated in a round chamber (illustrated in Figures 10-8b and 10-10) where the soil was loaded by compressed air (Lazebnik and Sivolap 1981). In this apparatus the soil could also be loaded by hydraulic press via a very stiff steel round plate. The lateral pressure developed in this apparatus was measured by four SDKS-type pressure cells identical to those used in earlier tests. In this case, however, the soil pressure-bearing plate was made slightly curved and flush-mounted with the chamber's curved wall.

For experimental consistency, initially soils within the round chamber were

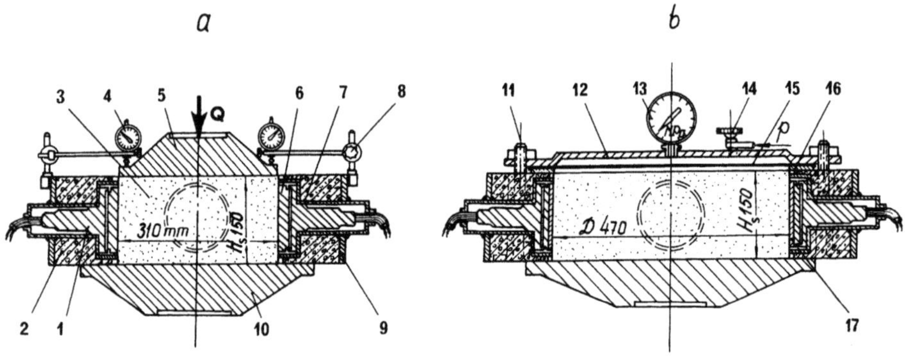

Figure 10-8. Devices for investigation of $K = f(\sigma_1/\sigma_z)$. (a) 310 × 310 × 150 mm chamber; (b) 470 × 470 × 150 mm chamber. 1. Soil pressure cell; 2. concrete; 3. soil; 4. dial pressure gage; 5. metal cover plate; 6. contact pressure cell; 7. cell socket; 8. gage holder; 9. steel tube; 10. steel base; 11. anchor bolt; 12. steel lid; 13. dial pressure gage; 14. air valve; 15. flexible diaphragm; 16. pressure ring; 17. curved metal strap attached to contact membrane flush with chamber wall.

loaded by the hydraulic press via a stiff plate. In this series of tests, the medium sand, in both loose and dense states similar to those used in the earlier tests conducted in the square-shaped apparatus, was tested. The results of these tests are shown in Figure 10-12. Data from earlier tests are added for comparison.

From Figure 10-12 one can see that the values of K measured in both devices differ from each other. For example, K for dense sand determined from tests in the square chamber is approximately 20% less than that measured in the round chamber, and the value of K obtained in the square chamber for loose sand is approximately 20% larger than that obtained in the round chamber. This confirms that the shape of the chamber has a profound effect on pressure measurements. To further clarify the issue, the soil was tested in the round-shaped chamber by applying loads using compressed air; the air pressure was applied to the sand surface via a flexible rubber diaphragm. The results of these tests and comparison with previous results are shown in Figure 10-13. From this figure it is seen that in dense sand the K value obtained by applying loads via a stiff plate and by compressed air is approximately equal to 1.33 at $\sigma_z = 0.1$ MPa and decreases to approximately 1.28 with an increase in pressure to $\sigma_z = 0.5$ MPa. This can be explained by the nature of the stiff plate/dense foundation soil interaction, which is typically characterized by stress concentrations in the areas located close to the plate's edges. As the load increases this area has the tendency to move slightly inwards from the edges toward the plate's center. On the other hand, loads applied by compressed air via the flexible diaphragm produce no such effects.

The obtained results for a loose sand (Figure 10-13b) are very different from those obtained in a dense sand (Figure 10-13a). For a loose sand, $K = f(\sigma_l/\sigma_z)$

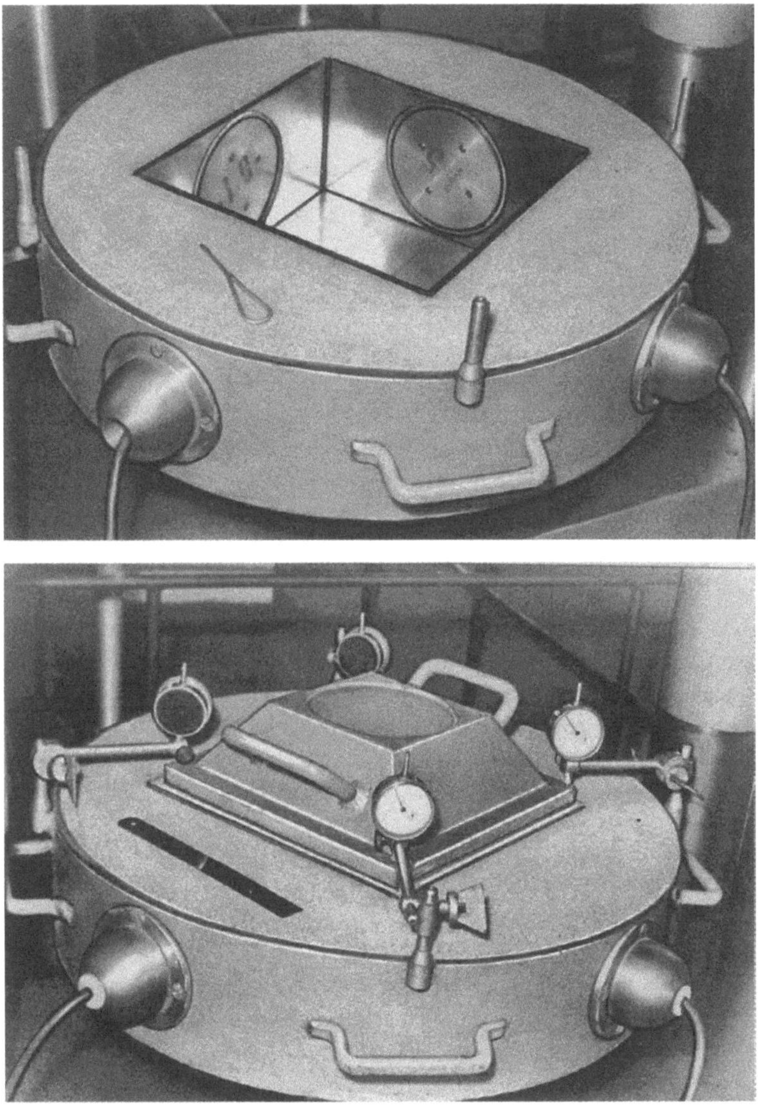

Figure 10-9. General view of apparatus for testing $K = f(\sigma_r/\sigma_z)$ shown in Figure 10-8a.

is almost unaffected by the method of load application (via a stiff plate or by compressed air via a flexible diaphragm).

The last problem investigated in this series of experiments was the reasons for larger values of $K = f(\sigma_r/\sigma_z)$ obtained in a square chamber for loose sand. These experiments were investigated in a special series of tests carried out in a square chamber (Figure 10-8a) by using eight miniature pressure cells arranged

Figure 10-10. General view of apparatus depicted in Figure 10-8b.

at two levels (four at each level) in a mass of soil, as indicated in Figure 10-14. To better determine the stress distribution within the soil mass, the instruments were located at different distances, *R,* from the chamber walls in this series of tests.

To exclude the effects of adjacent instruments on each other's performance, the distance between instruments was equal to but no less than 2*d,* with the exception of the central zone (here d_1 is cell diameter). The system experienced

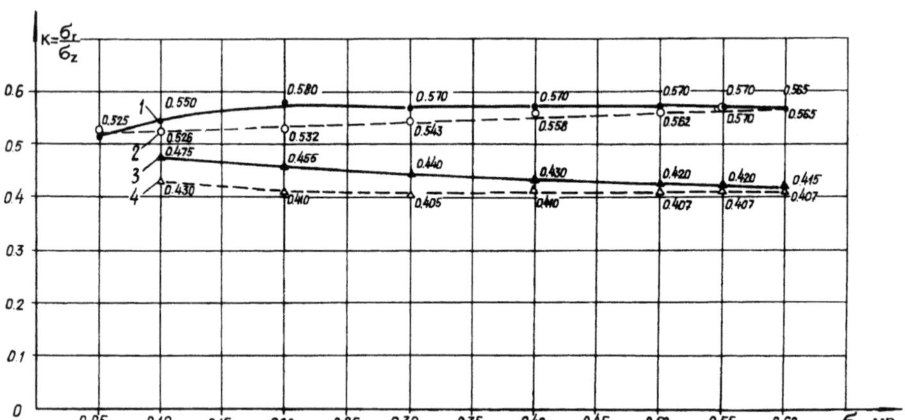

Figure 10-11. Results of experiments on determining $K = f(\sigma_r/\sigma_z)$ in 310 × 310 × 150 mm square device. 1 and 2. In medium sand, dense and loose, respectively; 3 and 4. in fine crushed stone, dense and loose, respectively.

Table 10-2. Grain Size Distribution and Other Parameters of Fine Crushed Stone

Quantity of Particles, mm, Percent by Weight							Density of Dry Soil (kN/m^3)		Void Ratio (e)		Angle of Response (deg)	
10–7	7–5	5–3	3–2	2–1	1–0.5	0.5–0.1	Loose	Dense	Max	Min	Loose	Dense
8.3	14.9	30.3	22.2	19.0	4.4	0.9	14.6	16.95	0.815	0.563	39	42.5

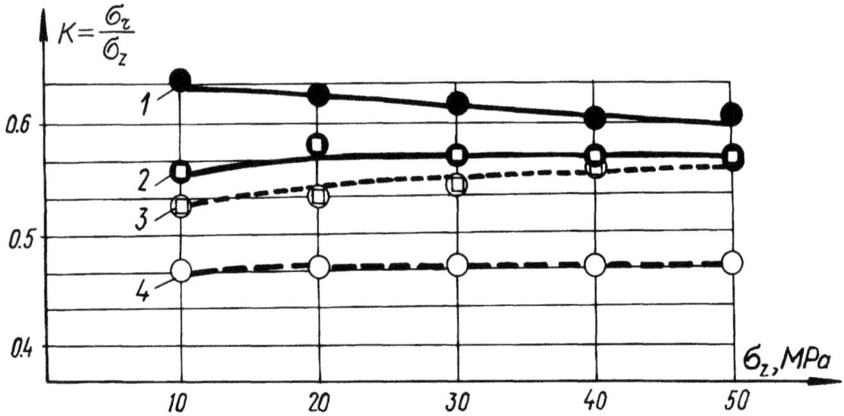

Figure 10-12. Comparison of soil lateral pressure coefficient, *K*, obtained from tests in a round-shaped apparatus (Figure 10-8b) (curves 1 and 4) and in square-shaped device (Figure 10-8a) (curves 2 and 3). 1 and 2. In dense sand; 3 and 4. in loose sand.

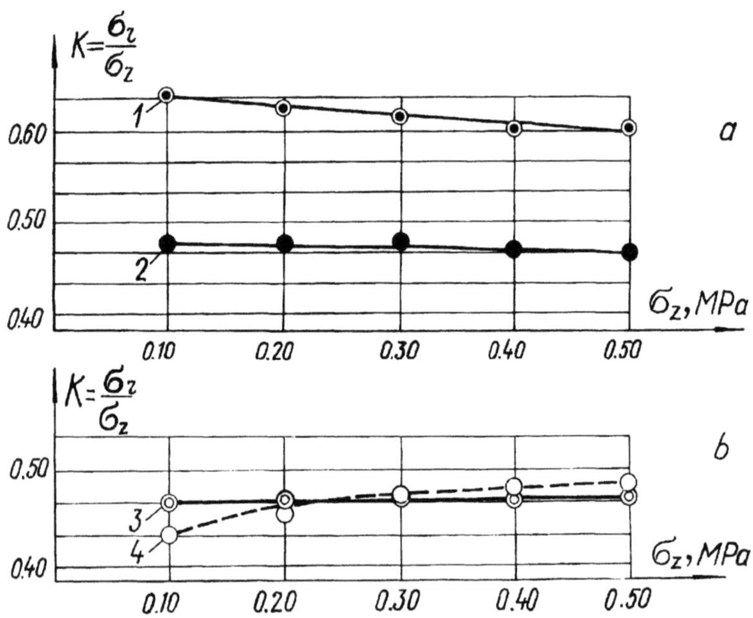

Figure 10-13. Coefficient of lateral pressure, *K*, of medium sand depending on the stiffness of loading surface in round-shaped test apparatus (Figure 10-8b). (a) In dense sand; (b) in loose sand. 1 and 3. Load applied via stiff steel cover plate; 2 and 4. pressure by compressed air applied via flexible diaphragm.

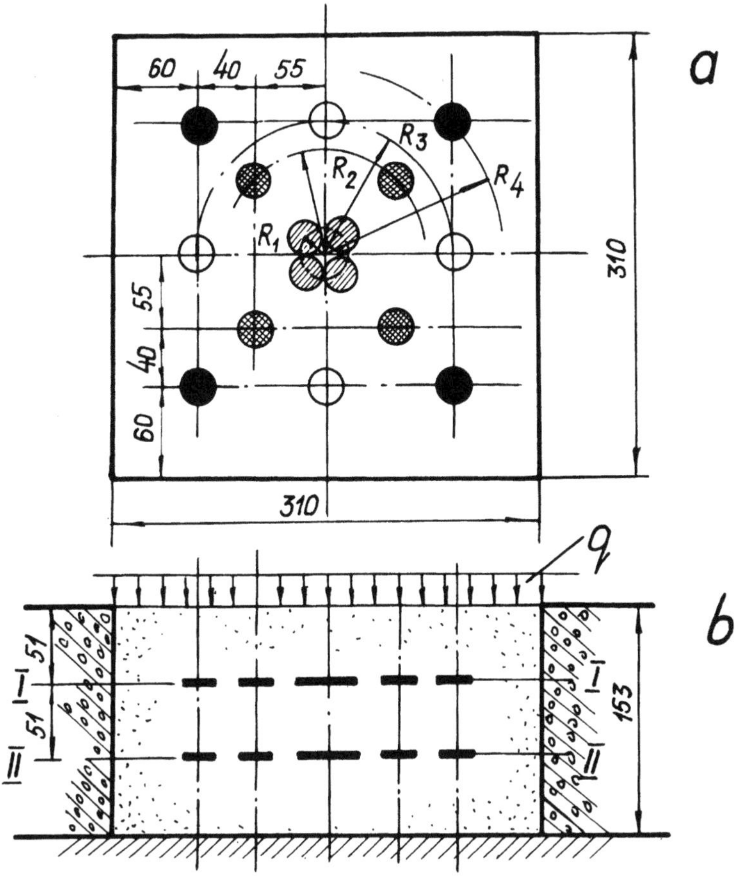

Figure 10-14. Plan (a) and cross section (b) of SMLP-1 setup with layout of soil pressure cells.

pressure loads equal to 0.2 and 0.4 MPa; the loads were applied via a stiff steel plate. All tests were duplicated three times; the instruments were rearranged each time, as indicated in Figures 10-14 and 10-15. The results of these tests are provided in Table 10-3.

From the data provided in this table it is seen that in a mass of loose sand the greatest reaction pressure exists in the central part of the chamber, and the smallest values of this pressure are registered in close proximity to the chamber corners (R_4 = 134 mm). This leads to the conclusion that in a square-shaped chamber the loose soil is stressed substantially differently than in a round-shaped device. On the other hand, for dense soils the stresses have greater values along the perimeter of the chamber, which causes stress concentrations in the corners of the device, with smaller pressures registered at its center. Because of this the

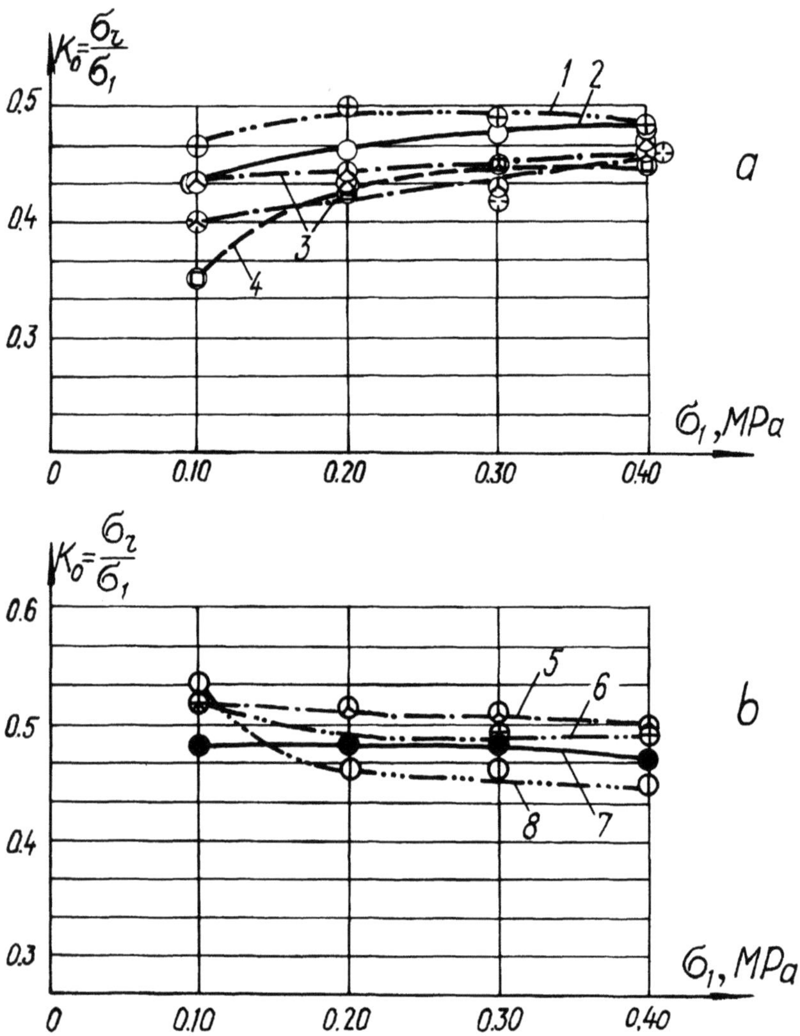

Figure 10-15. Results of investigations of coefficient of lateral pressure K_0 (K) of medium loose (a) and dense (b) sand in LCC and SMLP-2 setups as function of vertically applied stresses $\sigma_1(\sigma_z)$. 1, 3, 4, 5, 6, and 8. In LCC with soil pressure cells arranged in form of cross, star, and square, and when only one cell is used; 2 and 7. in SMLP-2.

contact pressure cells that are installed in the middle part of the chamber register smaller values of soil pressures than the actual average. Conversely, because in a loose soil the soil reaction concentrates in the middle part of the square chamber, the measured values of lateral pressures are smaller and in practice approach those measured in dense soil.

Table 10-3. Results of Vertical Stress Measurements in Mass of Loose Medium Sand Conducted in Square Device Illustrated in Figures 10-8a and 10-9

Level	Applied Pressure (MPa)	Measured Pressures (MPa) at R			
		$R_1 = 30$ mm	$R_2 = 78$ mm	$R_3 = 95$ mm	$R_4 = 134$ mm
I-I	0.2	0.226	0.238	0.241	0.209
	0.4	0.500	0.468	0.469	0.458
II-II	0.2	0.237	0.229	0.214	0.183
	0.4	0.528	0.463	0.426	0.387

Table 10-4. Values of Coefficients K_0 *and* K *in Medium Sand, Dense and Loose*

Pressure $\sigma_1(\sigma_z)$, MPa	Values of Lateral Pressure Coefficients				In Percent, Taking Value of K_0 as 100%	
	K_0		K			
	Dense	Loose	Dense	Loose	Dense	Loose
0.1	0.536	0.40	0.477	0.433	89	108
0.2	0.521	0.45	0.473	0.453	91	101
0.3	0.508	0.477	0.477	0.473	94	99
0.4	0.506	0.49	0.467	0.48	92	98

Table 10-5. Characteristics of Medium Sand Used in Experiments for Calculation of K_0 *Equation (10-1)*

Unit Weight of Dry Soil ρ_d in Experiments (T/m^3)		Angle of Internal Friction			Coefficients	
Max	Min	ϕ	tan ϕ	K_A	Sphericity δ	Roundness α
1.80	1.57	33	0.649	0.295	0.83	0.3

Table 10-6. Comparison of Experimental Values of K_0 *and Those Obtained by Equation (10-1)*

Values of K_0 Found in Experiments		Values of K_0 Calculated by Equation (10-1)		In Percent, Taking Experimental Data as 100%	
Dense	Loose	Dense	Loose	Dense	Loose
0.506	0.491	0.548	0.554	108.3	112.8

10.4 Comparison of K_0 with K and Proposed Formula for Determining K_0

Under the practically ideal confined soil conditions that exist in the calibration chambers and devices described in this work (Chapter 3) and used for determining values of the coefficient of soil lateral pressure, the principal stress σ_1 and actual vertical stress σ_z are practically equal ($\sigma_1 \approx \sigma_z$). Because of this it is interesting to compare the values of coefficients K_0 and K obtained in calibration chambers LCC and SCC described in Chapter 3, and in the round-shaped apparatus described earlier in this chapter that was used for the investigation of $K = f(\sigma_r/\sigma_z)$.

The relevant data is seeing in Figure 10-15 and also presented in Table 10-4. From these data it is seen that basically differences between K_0 and K are observed at vertical pressures $\sigma_1(\sigma_z)$ equal to 0.2 MPa.

As pointed out in Section 10.1, several workers have recommended expressions to calculate the value of K_0. Almost all of these expressions include ϕ, which is the angle of internal friction of the soil. Bishop (1958) suggests that there is a direct link between the coefficient of soil active pressure K_A and K_0. Based on Bishop's assumption of $K_A = \tan^2(45°-\phi/2)$, the following expression is proposed for determining the K_0 value (Lazebnik and Sivolap, 1974).

$$K_0 = K_A(\rho_s tan\phi + \alpha/\rho_d) \qquad (10\text{-}1)$$

where

ρ_s = unit weight of solid grains of sand; $\rho_s \simeq 2.65$ tonnes/m^3

ρ_d = density of dry soil in tonnes/m^3

α = dimensionless parameters that characterize roundness and sphericity of sand particles; given in Kolbuszewski (1958).

The data presented in Tables 10-5 and 10-6 show that the K_0 values obtained from equation (10-1) satisfactorily approximate the K_0 value obtained from the described experiments.

11

Soil Pressure on Retaining Walls

11.1 Introduction

It is well known that the magnitude of soil pressure on a retaining structure depends on the physical properties of the soil, the geometry of the contact surface, the character of displacement (deformation) of the retaining structure under soil pressure, and other effects, e.g., friction between the soil and the structure. Wall movement is a major contributor to the magnitude and distribution of soil pressure along the height of a wall.

Since Darwin's (1883) observations relating the lateral pressure of sand with movement of the retaining wall, hundreds of subsequent investigations have been carried out to confirm Darwin's model tests. Terzaghi (1934) observed that for submerged compacted sand the coefficient of lateral earth pressure decreased from about 0.6 at no yield to about 0.13 at $0.0005H$ forward wall movement (here H is the height of the wall). However, after less than two days of rest, this coefficient increased to 0.27.

Numerous laboratory and full-scale tests conducted in the past 50 to 60 years worldwide have indicated that in a great many cases the soil pressure redistributes along the contact surface area depends on wall deformations (movements). This is particularly characteristic of "flexible" sheetpile walls (bulkheads). Efforts to establish a realistic soil pressure value and its redistribution along the sheeting goes back a long time in the history of engineering arts and science. For a detailed discussion on soil/structure interaction and a comprehensive list of references, the reader is referred to Tsinker (1997).

In relationship to wall movement, lateral soil pressure is usually defined as at rest, active or passive. As discussed in Chapter 10, pressure at rest is assumed where no deformations or displacements of a retaining structure occur, and it is usually applied to the gravity-type quay walls built on a bedrock foundation. Active or passive soil pressures exist where the relative displacement between soil and structure (the structure's elements) may cause the soil either to expand

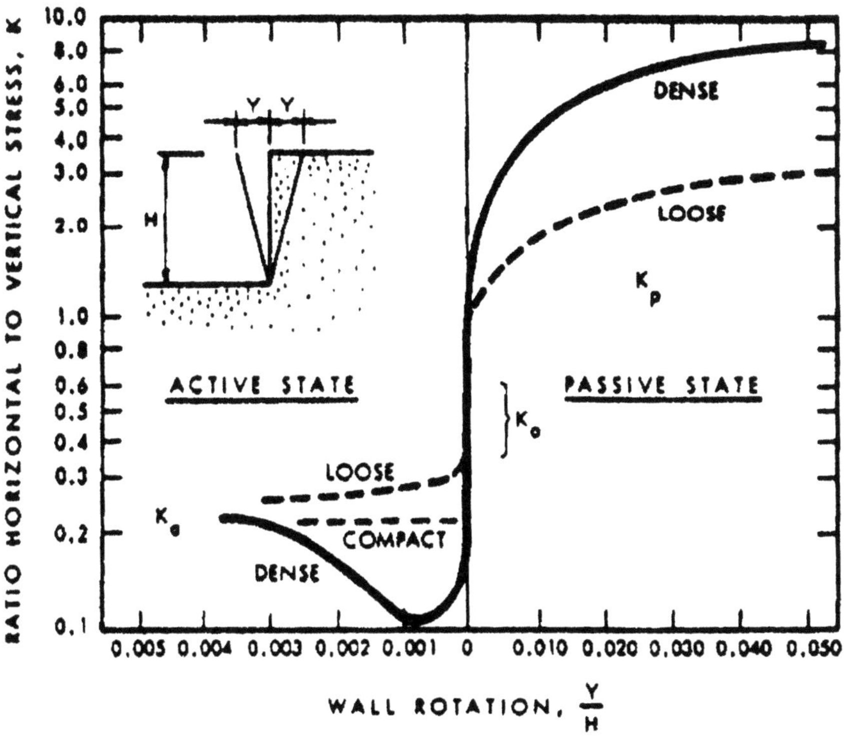

Figure 11-1. Effect of wall movement on earth pressure in sand.

(active state) or to contract (passive state). It usually applies to any type of soil-retaining structures built on soft foundation material. Any structural tilt produces lateral strain in the soil and alters the soil's horizontal stress conditions. Depending on the magnitude and direction of the tilt or deflection, the final horizontal stress can lie anywhere between the two limiting conditions: active and passive. The effect of wall tilt or deflection on the lateral pressure of sand is shown in Figure 11-1. For practical situations, the magnitude of wall tilt (*Y*/*H*) required to reach soil failure (active or passive) conditions in various soil types is indicated in Table 11-1.

The following sections discuss two important investigations on a flexible sheetpile bulkhead and one on a practically unyielding retaining wall conducted by this writer.

11.2 Large-Scale Investigation of a Single-Anchored Sheetpile Wall

Soil/flexible wall interaction is perhaps the most complex problem in the field of civil/geotechnical engineering. Rowe (1952) pointed out that the bending

Table 11-1. Magnitude of Wall Tilt to Achieve Active or Passive Soil Pressures

	Wall Tilt (Y/H)[a]	
Soil Type and Condition	Active	Passive
Dense cohesionless	0.001	0.02
Loose cohesionless	0.004	0.06
Stiff cohesive	0.01	0.02
Soft cohesive	0.02	0.04

[a]Y = horizontal displacement; H = height of wall.

moment in a sheetpile wall depends on at least 15 variables, including the stiffness of the wall, number and arrangement of wall supporting anchors, their yield under load, depth of the wall and its penetration into the foundation soil, foundation and backfill soil densities, creep parameters, miscellaneous soil parameters (e.g., angle of internal friction, cohesion), method of pile driving, wall constriction method, load character (e.g., uniformly distributed or concentrated), and others.

Sheetpile walls have been extensively investigated in the past 50 to 60 years; there are many published reports in the literature. Nevertheless, because of the extreme complexity of the problem, many aspects of sheetpile wall/soil interaction still need to be investigated.

In the 1960s this writer carried out comprehensive large-scale experiments in order to investigate a single-anchored sheetpile wall designed for construction in ports located on the Dnieper River. These investigations were performed in a specially designed bin 5 × 2.5 m in plan and 1.65 m high. Five 1.6-m-high models of sheetpile walls were tested in this bin. Each model consisted of 10 tongued and grooved sheetpiles. This prevented the potential formation of horizontal arches in the backfill soil, which may occur when a one-piece sheet is used to model the sheetpile wall. Cross sections of two types of sheetpile models used in this investigation are shown in Figure 11-2. The maximum free height of the tested wall is 1.3 m. Therefore, the maximum free wall height-to-bin width ratio is very close to 2. In practice, this makes it possible to ignore the effects of wall friction on sheetpiles located in the middle section of the model wall. Hence, measuring instruments were arranged and installed on the piles located in this area.

The sheetpiles were made from an aluminum alloy having a yield stress value

Figure 11-2. Cross Sections of Sheetpile Models.

F_y = 80 MPa. The wall's relative stiffness, St_r (a dimensionless number), was determined by a modified Rowe (1952) formula:

$$St_r = (H^4/EI)\rho b \qquad (11\text{-}1)$$

where

H = height (length) of the sheetpile
E = modulus of elasticity of pile material
I = pile's cross-sectional moment of inertia
ρ = soil density
b = unit width of the wall.

Values of I and St_r of the sheetpiles used in this investigation are given in Table 11-2. Both the foundation and backfill soils were medium quartz sand having characteristics similar to those noted elsewhere in this work.

The model setup system allows for the controlled yield of the sheeting in a range of 0.02 to 0.9% of H and for the accurate measurement of anchor forces. The model provides for 12 different attachment points for the anchor tie rod starting 50 mm from the wall top. By moving the position of the anchor up and down the wall height, it is possible to change the sheetpiling's relative stiffness for the same wall height.

A cross section of the instrumented sheetpile wall is depicted in Figure 11-3. The counterweight (1) was added to the system in order to balance the weight of the wale (9). A surcharge load to the wall was applied by water-filled containers with flexible bottoms. Sheetpiles were installed into the foundation sand by a driving method. Detailed discussions on sheetpile model construction and operation are given in Lazebnik (1961a, b, c).

The investigation of sheetpile wall/soil interaction was carried out in two stages. The first stage included model design and development of the investigation technique and criteria, as well as measurement of bending moments, reaction forces, and deflection of the sheetpile wall model. At this stage the soil pressure in both active and passive zones was determined indirectly from the measured diagrams of bending moments. Bending moments were measured by 20 pairs of bonded wire strain gages mounted along the model height. The strain gages were

Table 11-2. Values of Moment of Inertia I *and Relative Stiffness* St_r *of 1 Linear Meter of Sheetpile Wall Used in Experiments*

Sheetpile Model	1	2	3	4	5
Moment of inertia, I, (cm^4)	140	54	14	1.76	0.8
Relative stiffness, St_r	1.14	2.95	11.37	90.43	199

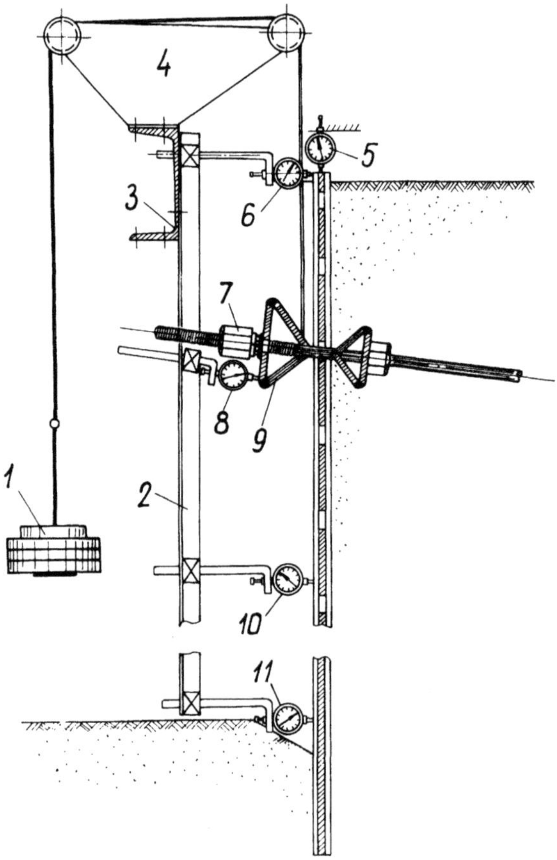

Figure 11-3. Typical cross section of instrumented sheetpile wall. 1. Counterweight; 2. steel angle to support dial gages; 3. steel channel; 4. steel plate; 5. dial gage for measuring wall vertical displacement; 6, 10, and 11. dial gages for measuring wall horizontal displacements; 7. nut; 8. dial gage for measuring yield of anchor tie rod; 9. steel wale.

protected from damage by mechanical and other effects; they were calibrated on a special stand before the piles were installed. The calibration stand is shown in Figure 11-4. The sheet pile relative deflections were measured by a mirror galvanometer with a relative accuracy of 1.5×10^{-6}. Furthermore, sheetpile deflections, vertical settlements, and anchor rod elongations were measured by standard dial gages with an accuracy of 0.01 mm.

During the second stage the soil pressure against the sheetpile wall was measured directly by specially designed small pressure cells (Chernyshova 1963). These cells are described in Chapter 4 and illustrated in Figures 4-4 and 4-5. Complete sketches of the instrumented sheetpiles and the location of instruments and gages are shown in Figures 11-5 and 11-6.

Figure 11-4. Calibration of sheetpile model—working moment.

The results of measurements of bending moments and related deflections for three sheetpile models having different relative stiffnesses, St_r, are shown in Figure 11-7. In this figure the bending moment diagrams obtained from the tests are compared with the theoretical diagram calculated by the conventional approach (examples are found in Tsinker 1997) based on the assumed wall/soil friction angle, δ, equal to zero in the active zone and equal to $2/3\phi$ in the passive zone, where ϕ is the soil angle of internal friction. In these experiments the model was constructed in the following sequence. First the sheetpiles were driven into the foundation soil to a depth of 0.6 m and anchored. This was followed by installation of the backfill, which was placed in layers with a slight ramming to achieve the desired density.

From Figure 11-7 it follows that the stiffer the wall, the greater a positive bending moment is developed in it. At the same time, the value of the anchor force, Δ_p, decreases with an increase in wall flexibility. It should be noted that the maximum measured yield of the wall at the tie rod level was less than 1.0 mm. According to this series of tests, the bending moment in the sheetpile wall changes its sign just above the dredge line, and the lowest value of the negative bending moment develops in the pile having the highest St_r value. This latter pile also has the smallest active span, the distance between the anchor point and the pile fixity point in the founding soil.

Soil pressure diagrams for sheetpile models 2 and 3 (see Table 11-2) obtained by a double differentiation of the relevant diagrams of bending moments are presented in Figure 11-8. For comparison, the conventional soil pressure diagrams of the active and passive pressures obtained by classical soil pressure theory

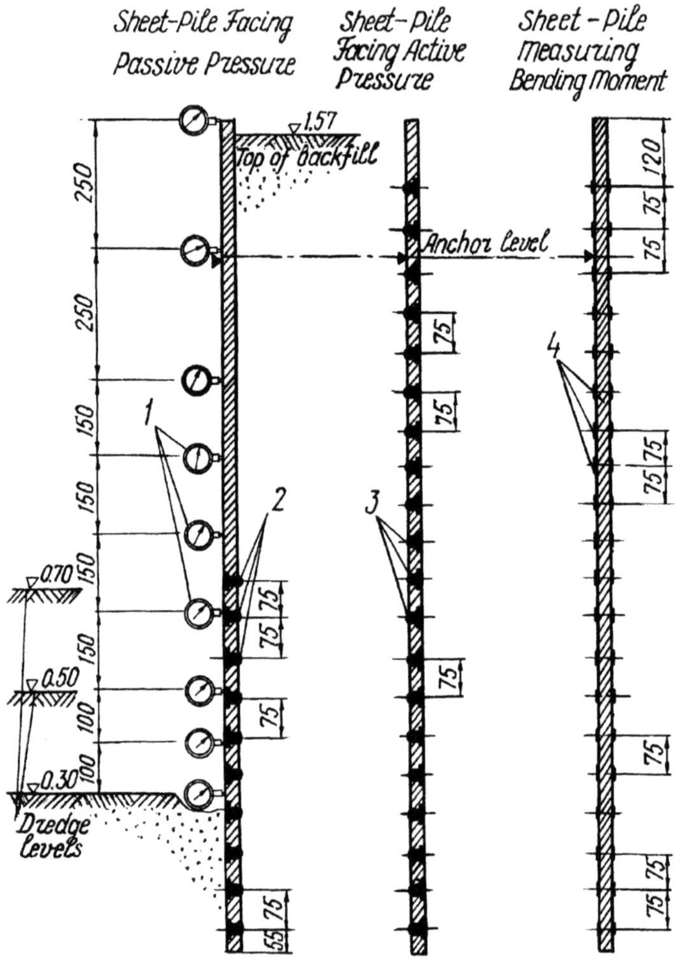

Figure 11-5. Typical cross sections of instrumented sheetpile. 1. Dial gage; 2. contact pressure cells for measuring passive pressures; 3. contact pressure cells for measuring active pressures; 4. pairs of strain gages for measuring bending moments.

(conventional method) for different values of δ are also shown in this figure. As is seen in the figure, the shapes of the soil pressure diagrams in both the active and passive zones differ substantially from those derived from the classical soil pressure theory. Again, it should be noted that the conventional theory of sheetpile wall analysis does not account for such important factors as wall deflection, stiffness, construction method, and sequence, etc. The experimental soil pressure diagrams shown in Figure 11-8 indicate an increase in the soil pressure at the anchor support and in the area close to the dredge line zone, and a decrease in soil pressure in the midspan area.

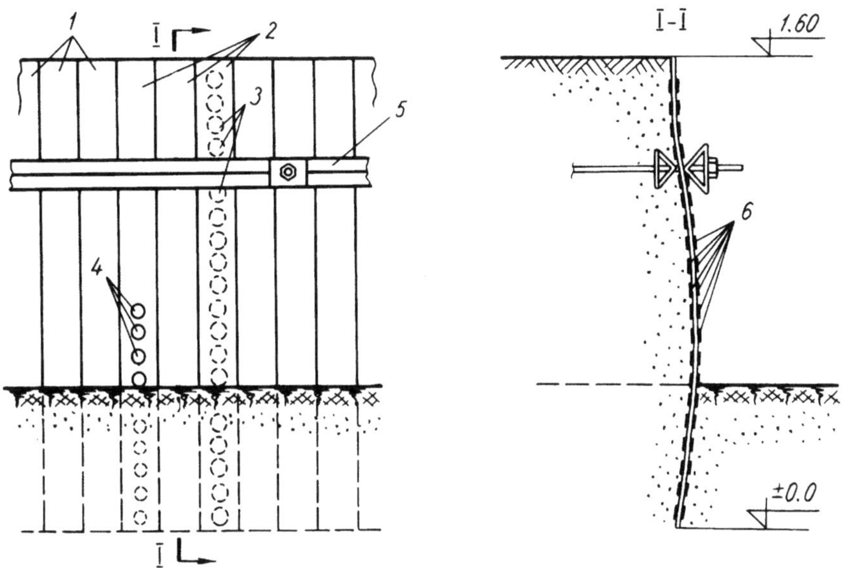

Figure 11-6. Relative positions of instrumented sheetpiles—elevation and cross section. 1. Regular sheetpiles; 2. instrumented sheetpiles, 3. soil pressure cells on active pressure side; 4. soil pressure cells on passive pressure side; 5. wale; 6. strain gages.

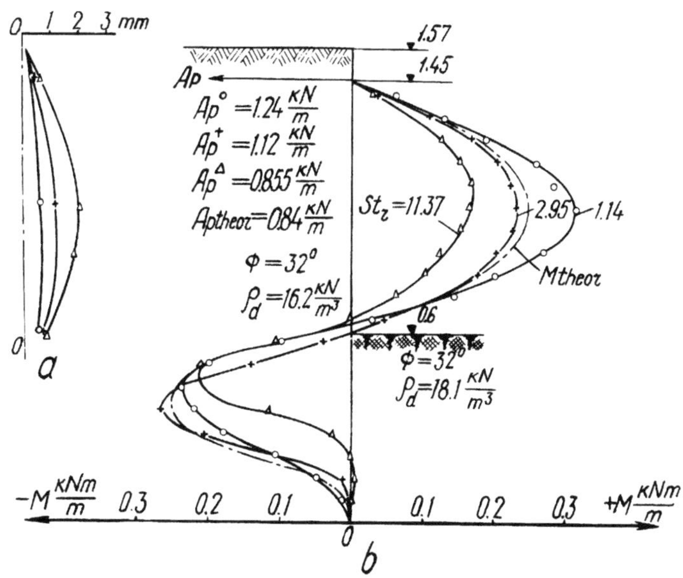

Figure 11-7. Performance of sheetpile walls with different relative stiffnesses (St_r). (a) Deflection profiles; (b) diagrams of bending moments.

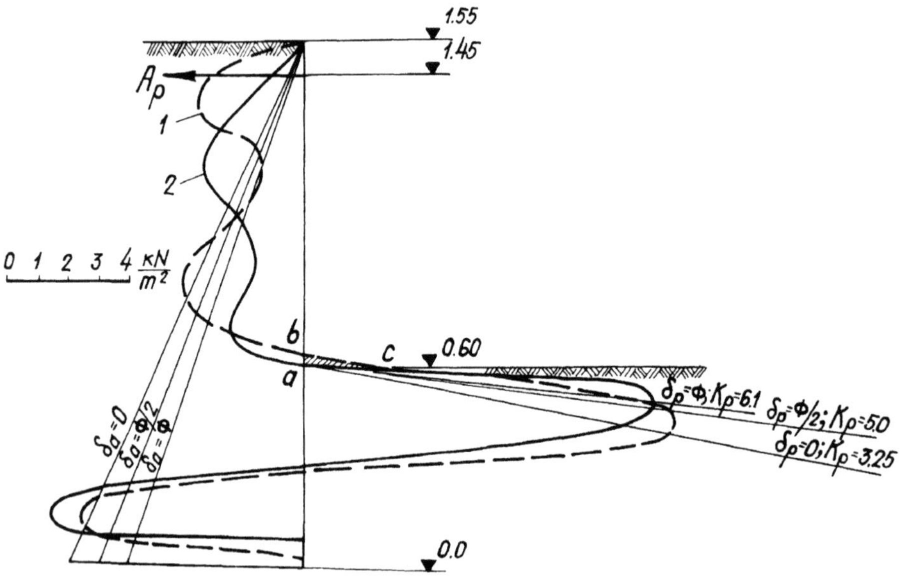

Figure 11-8. Lateral soil pressure diagrams obtained by double differentiation of bending moment curves. 1 and 2. For walls composed from sheetpiles 2 and 3, respectively (see Table 11-2).

Subsequent calculations based on soil pressure diagrams indicate that the shape (redistribution of soil pressure) of the soil pressure diagram in the active zone alone does not result in smaller values of bending moments in a relatively flexible sheetpile. The values also depend on the shape of the soil pressure diagram in the passive zone, which results in a shorter active span of the sheetpile and therefore a smaller value of the bending moment. Earlier this was indicated by Tschebotarioff (1948 and 1973), who perhaps first explained the phenomenon of substantial reduction in bending moments in flexible sheetpile walls.

The investigations by this writer confirmed the results obtained by Tschebotarioff. It was also found that for very stiff sheetpiles ($St_r = 1.14$) and nonyielding anchors the shape of the soil active pressure diagram closely resembles that characteristic of the soil pressure "at rest" obtained by classical theory at $K_0 = 0.55$; in this case the gravity center of the soil passive pressure diagram is usually located somewhat lower than that characteristic of the flexible sheetpiles. The combination of both the shape of the passive pressure diagram and the position of its gravity center results in a longer span and subsequently a greater value of the bending moment shown in Figure 11-7.

It should be pointed out that the double differentiation of the bending moment diagram is associated with a substantial error in determining the shape of a soil pressure diagram. This explains why the soil passive pressure shown in Figure

11-8 is projected above the dredge line (shaded area "abc"), which does not make sense.

Directly measured soil pressure diagrams were obtained during the second stage of the investigation. Figure 11-9 illustrates measured deflections, soil pressure distribution, bending moments, and calculated shear forces in the sheetpile wall made from model 2 (Table 11-2) driven 70 cm into the foundation soil. The piles were driven into a relatively dense medium sand and backfilled similar to the procedure in the previous set of tests. The data plotted in Figure 11-9 is a result of three practically identical tests. As in previous tests, the yield of the anchor did not exceed 0.1% of the total model height, H.

Because of deeper penetration in the foundation soil (0.7 m vs 0.6 m) in the latter set of tests, the wall reaction at the anchor point is smaller than in the earlier experiments (see Figure 11-7). As is seen from Figure 11-9, the readings of soil pressures in the wall active zone are substantially scattered. Despite this, however, it is possible to see an increase in the ordinates of soil pressure in the area of the anchor point and a decrease in the areas of wall maximum deflection that make up the wall's midspan, and at the model lower part.

The results of measurements of soil pressure in the passive zone produced a rather consistent set of readings. In this regard, the obtained soil pressure diagrams differ substantially from that calculated by classical theory. The values of the soil pressures and the shapes of the soil pressure diagrams are consistent with those obtained by Tschebotarioff (1948, 1973) and Rowe (1952). The shapes of the soil pressure diagrams to a great extent resemble those obtained by the theory of a beam on elastic half-space (Hetinyi 1946; Gorbunov-Passadov and Malikova 1973) or by finite element analysis. Under the conditions of the tests in question, the resultant of the passive pressure diagram is located at a distance approximately $2/3D$ from the lower end of the sheetpiling. Here D is the sheetpile depth of penetration.

The calculations indicate that the value of the bending moment obtained in these tests has a value approximately 43% smaller than that obtained by classical theory. The passive soil pressure diagrams obtained by direct measurements in soils of varying densities and at different depths of pile penetration are shown in Figure 11-10. As stated earlier, in the previously discussed test results the anchor yield, Δ_A, was equal to approximately 0.1% of H. Note that in practical cases it may be much larger. For example, Terzaghi (1934, 1953) suggests that Δ_A, in some cases, can reach up to 1% of H.

The effect of sheetpile wall displacement at the anchor point was investigated, and the results of this study are shown in Figure 11-11. From this figure, it follows that both the bending moments in the upper part of the wall and the anchor loads are reduced with an increase in wall yield, Δ_A, at the anchor point. At the same time this results in an increase in the bending moment in the lower part of the wall. As is seen from Figure 11-11, as the wall yields, the resulting changes in the values of bending moment and anchor force are basically a function

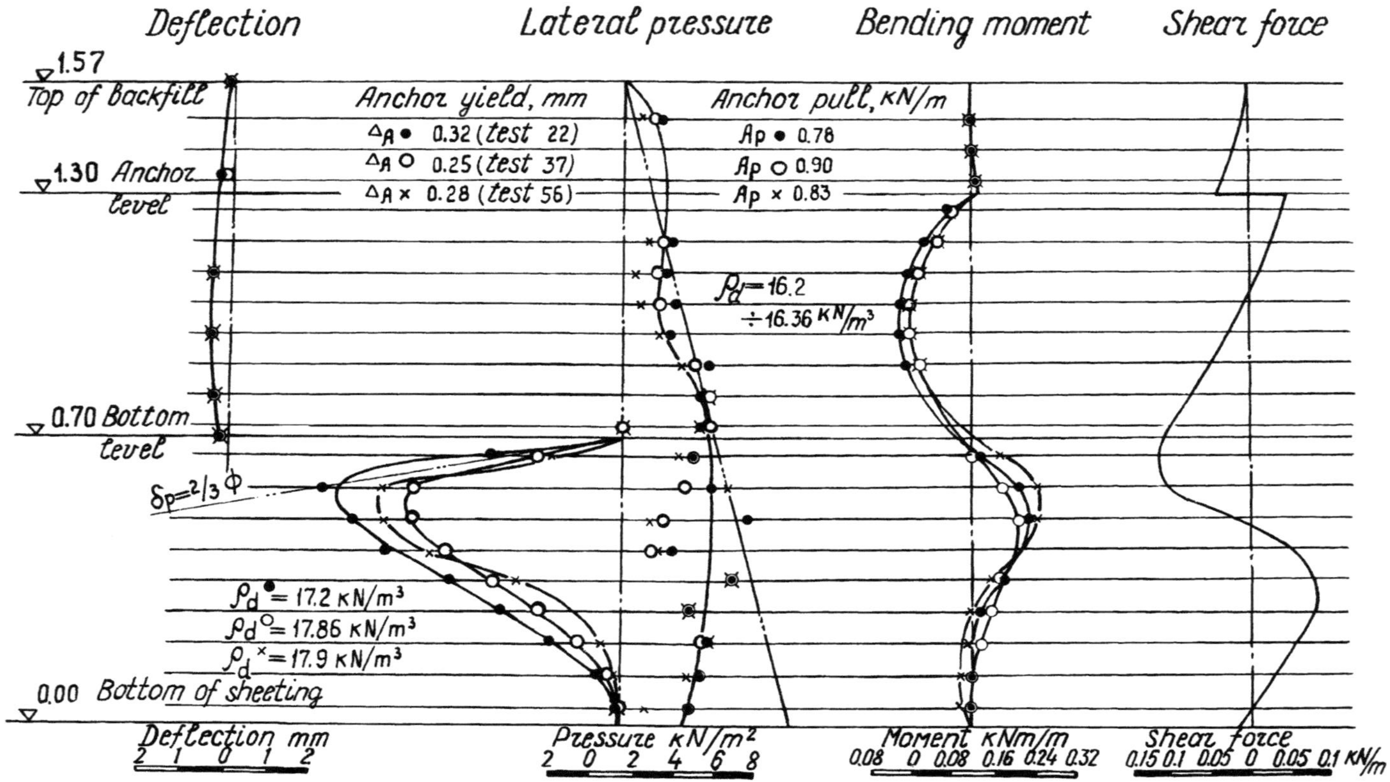

Figure 11-9. Performance of wall composed from sheetpiles model 2 (Table 11-2) driven in dense foundation soil.

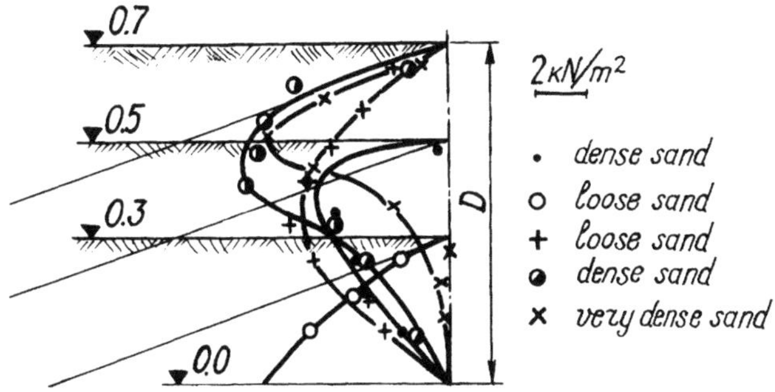

Figure 11-10. Diagrams of soil passive pressure obtained in experiments by direct measurements.

of wall stiffness and yield. It should be noted that in a stiffer wall (sheetpile model 1, Table 11-2) the bending moment in the lower part of the wall grows more intensively as the Δ_A increases. The basic reason for this is that under substantial yield at anchor point, the wall performs similarly to a cantilever beam fixed in a foundation soil; this is why the greater the wall stiffness, the more pronounced a cantilever effect is observed. The effects of wall yield on pressure distribution in the wall passive zone are shown in Figure 11-12. From this figure it is seen that in the denser foundation soils the passive pressure of soil is greater than in the less dense soils, which is quite expected.

It is possible to conclude that, in both medium density and dense foundation soils, with an increase in Δ_A the concentration of soil pressure decreases in the anchor zone, is practically not affected in midspan, and increases in the lower part of the wall. This results in a decrease in the value of the anchor force and bending moment in the wall midspan. The latter is a result of a decrease in the wall active span that is defined as the distance between the position of the resulting passive pressure force and the wall anchor point. Conversely, the moment at the point of wall fixity within the foundation soil increases.

A sheetpile wall can be constructed as backfilled as well as "dredged." The dredging construction usually proceeds in the following sequence: first, sheetpiles are driven; then the anchor system is constructed and the additional backfill material is placed up to the wall grade (if required); finally the soil on the wall face is dredged (excavated). Naturally, during these two construction procedures the sheetpile wall deflects differently, which results in a different soil pressure distribution along the sheeting. The latter has a pronounced impact on sheetpile wall performance (Tschebotarioff 1973). [For a detailed discussion on this subject, the reader is referred to Tsinker (1997).]

In general, it was found that the anchor force of the dredged bulkheads, depending on the depth of dredge, under the wall nonyield condition is 20 to 50% higher in comparison to that of the backfilled wall. At the same time bending moments in the wall span are much smaller than those in the backfilled walls.

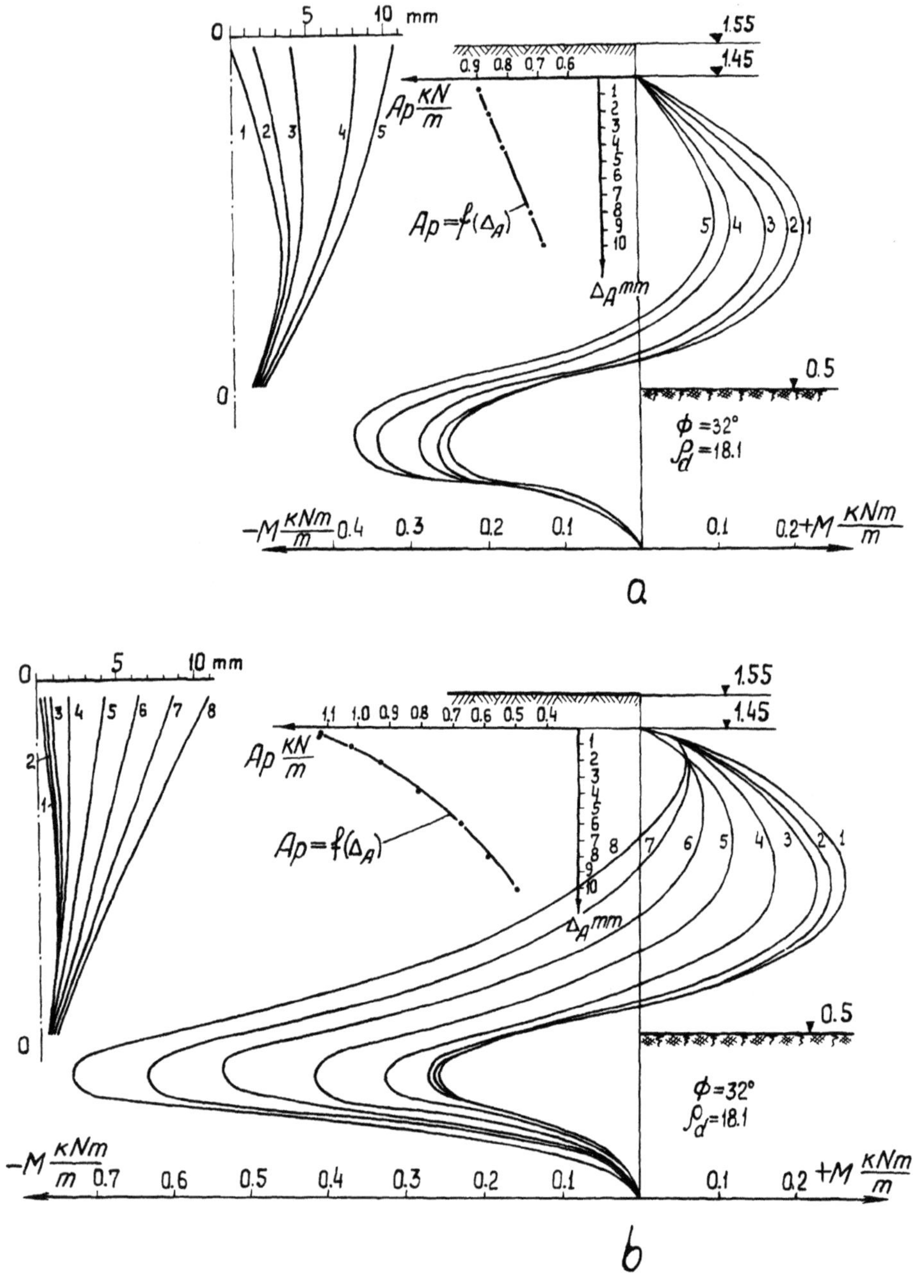

Figure 11-11. Bending moments and anchor forces as function of wall yield at anchor point. (a) Wall composed from model 3 (Table 11-2); (b) wall composed from model 2 (Table 11-2).

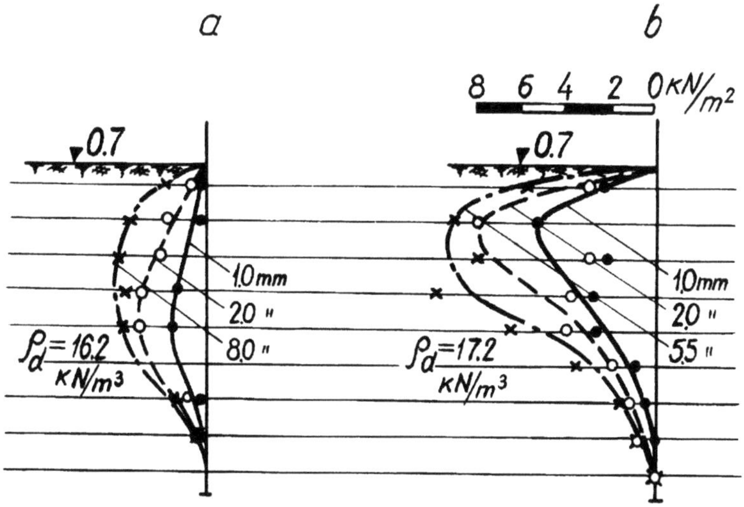

Figure 11-12. Passive soil pressures as a function of wall yield at anchor point (Δ_A = 1, 2, 5.5, and 8 mm). (a) Passive pressure distribution in loose sand (ρ_d = 16.2 kNm^3) (b) Passive pressure distribution in medium density sand (ρ_d = 17.2 kN/mm^3).

This was confirmed by 40 tests that were carried out on models with different stiffnesses, driven into foundation soils having different densities. An example is shown in Figure 11-13. In general, depending on the sheetpile stiffness, density of the foundation soil, and presence of a uniform surcharge load placed on a grade, the bending moment in the dredged wall was 15 to 30% smaller than in the backfilled walls.

In these experiments the increase in the anchor load was found to be less pronounced, basically because of the low height of the wall projected above the anchor point. In general, however, in dredged bulkheads, the upper part of the wall movement toward the backfill can cause the load concentration at the anchor point to be substantial. Besides, as pointed out by Tschebotarioff (1973), the anchor force of dredged bulkheads is increased by the formation of vertical arches in the backfill soil; these arches are supported at the anchor point and at the dredge level. Naturally, the lateral soil pressure at these points has increased values. The arches can be stable under the wall nonyielding condition; they may, however, partly or completely collapse when the wall yields. These findings confirmed test results reported by Rowe (1952).

The effects of a uniform surcharge load on wall performance were investigated in a special series of tests. The typical results of these tests are illustrated in Figure 11-14. Note that the bending moment, anchor reaction, and lateral soil pressure illustrated in this figure are due to action of the surcharge load only;

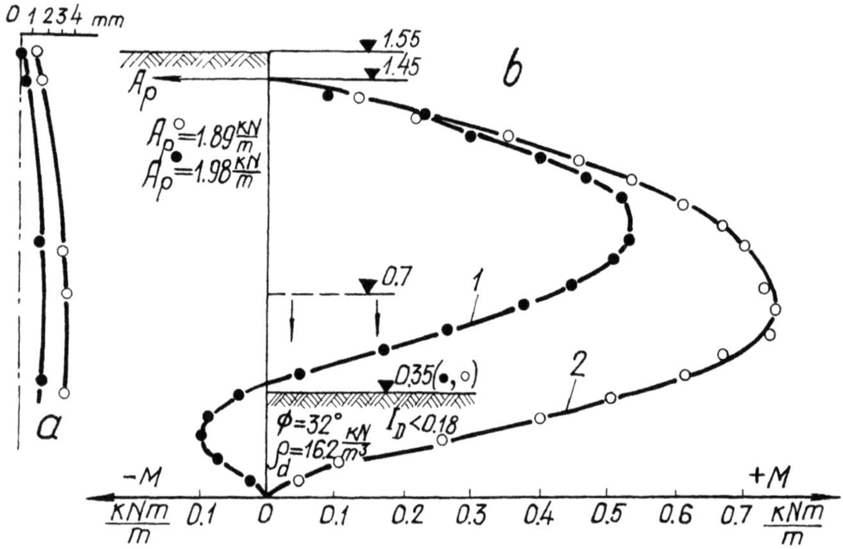

Figure 11-13. Typical example of soil/sheetpile wall interaction depending on sequence of construction. (a) Wall deflection; (b) bending moment diagrams. 1. Dredged bulkhead; 2. backfilled bulkhead.

they act in the wall system in addition to those effects resulting from the loading by the backfill material. The diagram of the lateral soil pressure depicted in Figure 11-14 differs substantially from the standard rectangular diagram obtained by classical theory. In the final analysis the surcharge load produces a smaller bending moment in the wall span in comparison with that obtained by classical theory, and the anchor load due to the uniform surcharge load does not exceed that obtained by classical theory. The obtained shape of the diagram of lateral pressure due to the uniform surcharge load is close to that obtained by the theory developed by Fröhlich (1934) and using the concentration coefficient $\nu = 5$.

In the experiments described in this section, the uniformly distributed load was applied in the area of the "active wedge" behind the wall that develops in the backfill material. Extension of this load beyond the "active wedge" resulted in an insignificant increase in the anchor force (3 to 6%) and the bending moment in the wall (4 to 7%).

11.3 Soil Pressure Against Nonyielding Retaining Wall

This wall is part of a pumphouse constructed on a foundation soil comprised mainly of fine sand having an angle of internal friction ϕ = 33 to 34°, cohesion $c = 0$, density ρ_d = 16.5 to 17 kN/m^3, and modulus of deformation E_s = 40 to

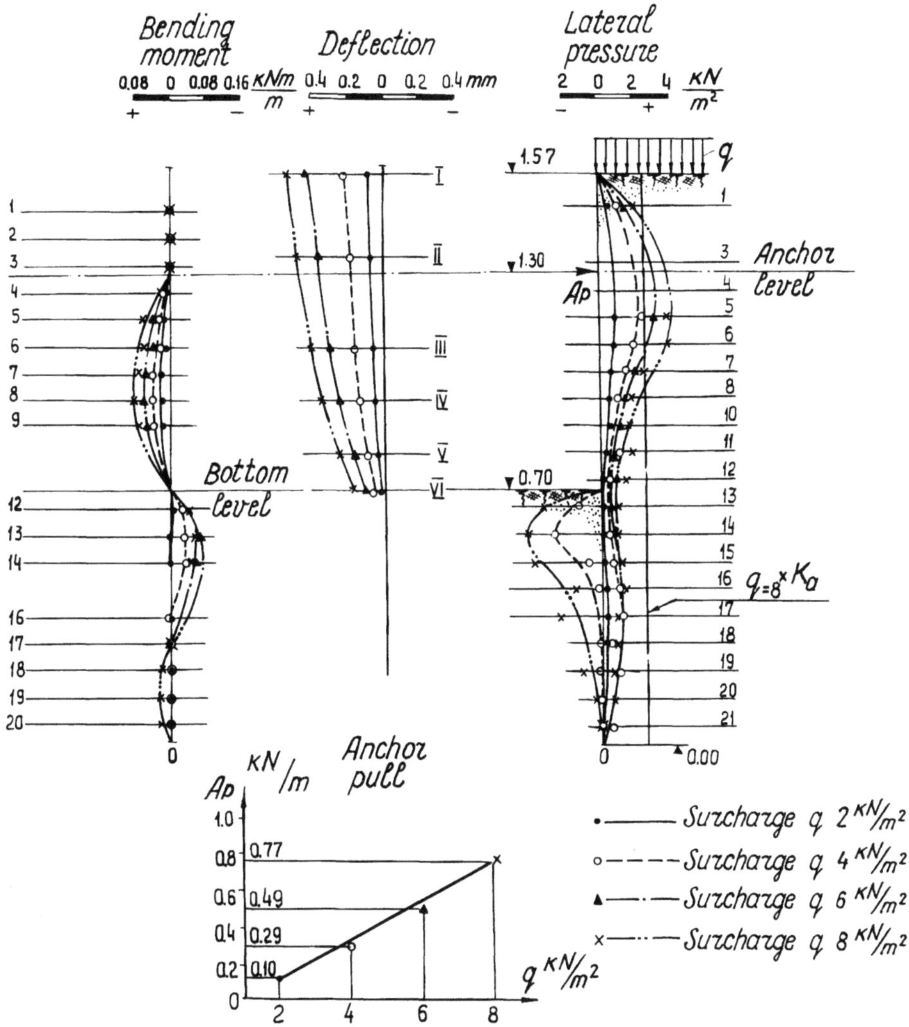

Figure 11-14. Effects of uniformly distributed load on bending moment, anchor force, and resulting lateral pressure on sheetpile wall as function of wall deflection.

45 MPa. The structure is practically nonyielding. A typical cross section of this pumphouse is depicted in Figure 11-15.

Backfill soil was placed in layers 0.5 to 0.7 m thick and densified by a bulldozer moving parallel to the wall. An average angle of internal friction of the backfill material is $\phi = 32°$, the density $\rho_d = 16$ kN/m^3, and the modulus of deformation

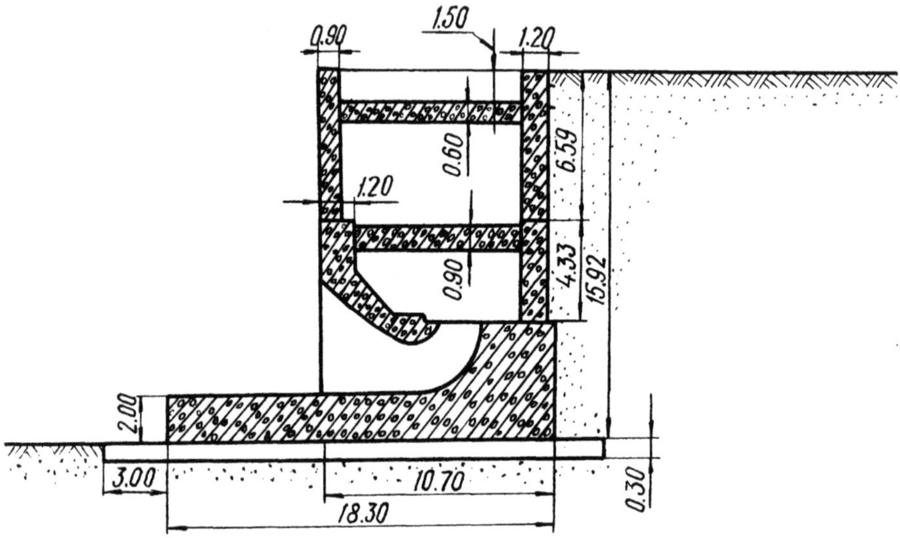

Figure 11-15. Investigated pumphouse. Typical cross section.

varied from E_s = 20 MPa in the upper part of the wall to E_s = 30 MPa in the lower part. The soil pressure on the wall was measured by SDKS-type pressure cells installed flush with the wall face exposed to soil pressure. The cells were installed at eight levels in three vertical sections of the wall. The instrument arrangement and the process of their installation are shown in Figures 11-16 and 11-17.

Soil pressure readings were taken as the wall backfilling progressed; e.g., when the depth of the backfill soil behind the wall was equal to 7, 9.4, and 15.92 m. All measurements were taken either almost immediately or within a relatively short period of time on completion of the backfilling to a predetermined elevation. Final soil pressure measurements were made 10 months after the backfill was completed. The results of the soil pressure measurements are shown in Figure 11-18. In this figure the measured soil pressure diagrams are compared to theoretical ones obtained by conventional classic calculations using different values of the coefficient of soil lateral pressure, *K*. From Figure 11-18 it is seen that at the first stage of wall backfilling (up to elevation –8.90 m), the shape of the soil pressure diagram (curve 1) is practically triangular. At this stage the coefficient of soil lateral pressure is equal to approximately 0.5 to 0.55. At elevation –6.5 m the shape of the soil pressure diagram started changing its configuration (curve 2), and at elevation zero the shape of the soil pressure diagram is close to triangular at its upper part with a characteristic reduction in the value of the soil pressure at the lower part of the wall (curve 3). The corresponding values of *K*

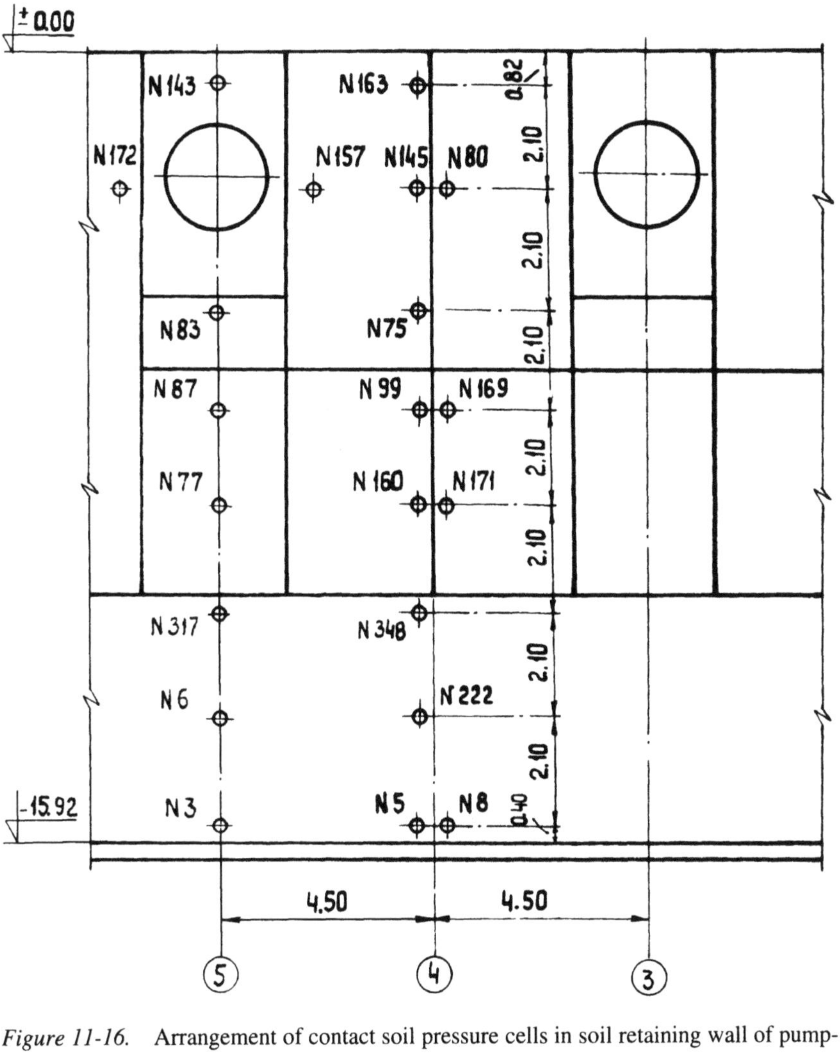

Figure 11-16. Arrangement of contact soil pressure cells in soil retaining wall of pump-house shown in Figure 11-15.

indicate that at the upper portion of the wall $K \simeq 0.5$, in the middle part it varies from approximately 0.4 to 0.44, and at the bottom part of the wall the K value is reduced greatly.

The measurements taken 10 months after the backfill was installed (curve 4) indicate that, despite the fact that the total soil pressure did not change much, the intensity of the soil pressure along the wall is substantially redistributed; it is substantially smaller at the upper part of the wall, heavily concentrated in its

Figure 11-17. Installation of contact pressure cells on wall of pumphouse shown in Figure 11-15.

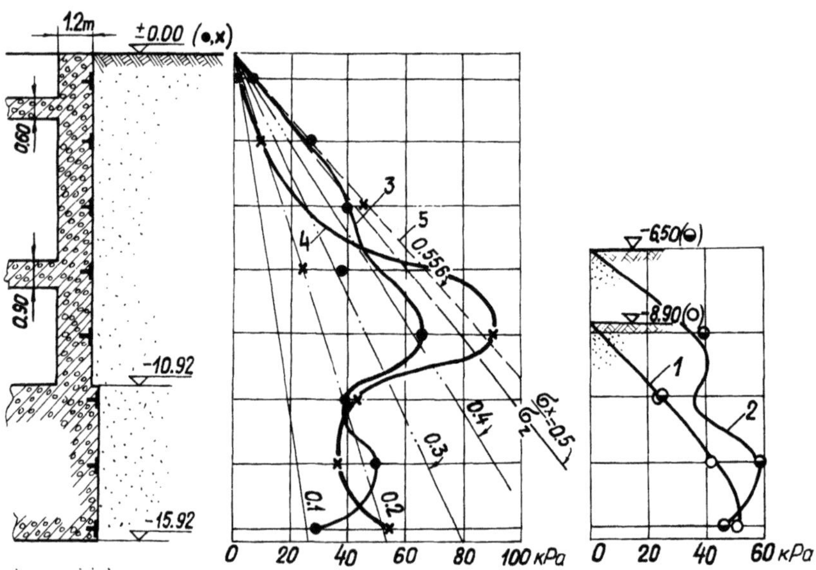

Figure 11-18. Measured diagrams of lateral soil pressure on a stiff wall of the pumphouse. 1. Backfill at elevation −8.9 m; 2. backfill at elevation −6.5 m; 3. at elevation 0.00; 4. 10 months later after completion of wall backfilling; 5. calculated soil pressure diagram using coefficient of lateral pressure K_0 = 0.556 obtained from equation (10-1).

middle section, and is not changed much at the lower part of the wall. It should be noted that the decrease in soil pressures on a rigid wall with time was observed by Muhs (1947) and Tschebotarioff (1973). Finally, we point out that in the upper part of the wall, the coefficient of lateral pressure calculated by equation (10-1) based on actual field data is equal to 0.556 (curve 5).

12

Vertical Distribution of Stresses Within Soil Mass at the Central Part of a Stiff Slab

The results of numerous investigations conducted in the past 30 to 40 years indicate that stress distribution by soils is much less pronounced than that obtained from calculations based on the theory of elastic soil/structure interaction, e.g., the Boussinesq solution. Some investigators, however, arrived at exactly the opposite conclusions (Turnbull et al. 1961; Abramov et al. 1968).

In the early 1970s this writer and co-workers carried out a series of tests on very stiff rectangular and round blocks founded on a river quartz sand and loam. As in all identical earlier described experiments, these blocks were loaded by hydraulic jack.

12.1 Stiff Rectangular Concrete Block 1.8 × 0.6 × 0.7 m

This block was tested on a sand foundation having prevailing particles of 0.5 to 0.25 mm (39.7%) and 0.25 to 0.1 mm (43.3%) in size, and on a loam (see Table 5-5 for the properties of this loam). The initial density of the sand was 16.6 to 16.9 kN/m^3 at a water content w = 1.6 to 2%, and the initial density of the loam was equal to 17.8 to 18.1 kN/m^3 at w = 10.2 to 10.6%. The load was applied in increments of 0.1, 0.2, and 0.3 MPa, and stresses in soil mass were measured at four levels, as indicated in Figure 12-1.

The pressure within the soil mass was measured by stiff flat membrane-type pressure cells (d = 50 mm) having a bonded-toil strain gage-type measuring system. For the sandy foundation, stresses were measured in two series of experiments; for the loam foundation, they were measured in one series. Each series of tests included two loading cycles. The stresses at elevations –0.4 and –0.8 m were calculated as the average readings of three instruments installed at each level. At elevations –0.1 and –1.2 m there was an average reading by 10 and 6 cells, respectively.

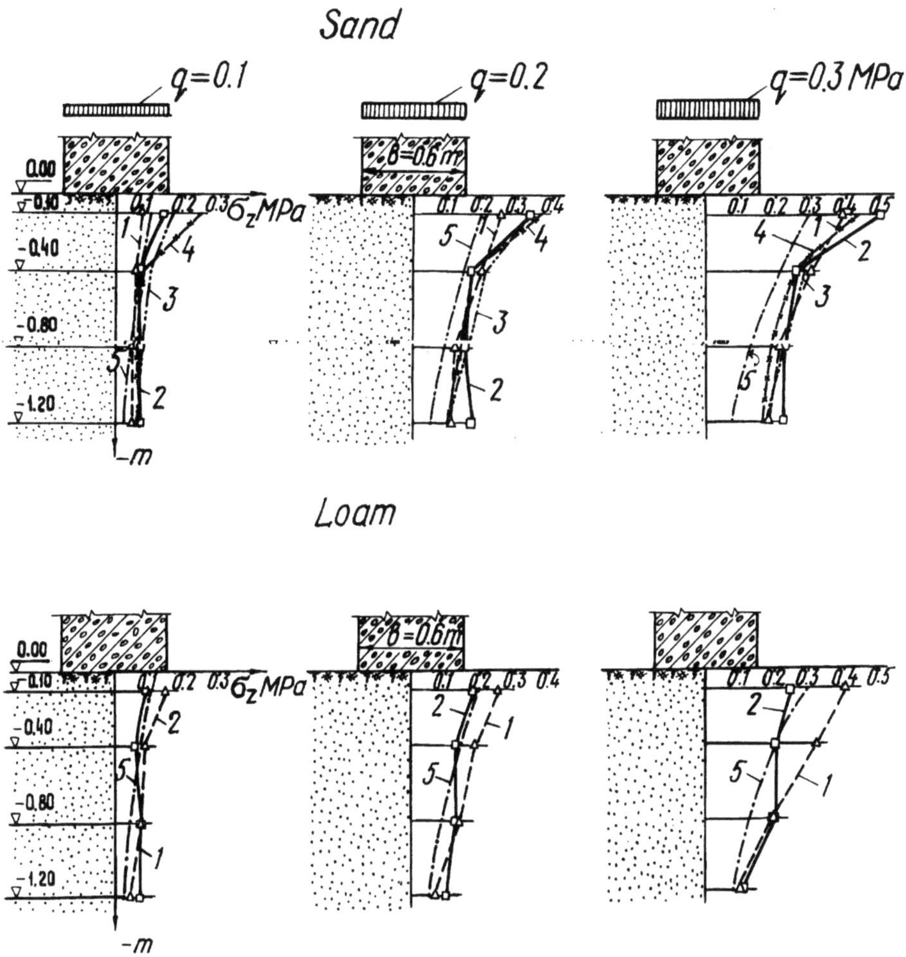

Figure 12-1. Soil pressure distribution in soil mass under stiff rectangular block 1.8 × 0.6 × 0.7 m. 1 and 3—stresses obtained in the first loading cycle (1 and 2 series of tests respectively); 2 and 4—same in the second loading cycle; 5.—stress distribution calculated from classical theory of elasticity.

The test results are presented in Figure 12-1. As is seen from this figure, at the first loading cycle, the measured stresses in mass of soil exceed the values of the stresses obtained by the theory of elasticity (Boussinesq 1885; Timoshenko 1934). In the second loading cycles this difference is much smaller. In these experiments substantial scatter in cell readings was present in the soil pressure measurements at elevation 0.1 m. This can be explained by the small diameter of the cells, but most importantly by too close a proximity of the cell active surface to the underside of the very stiff block.

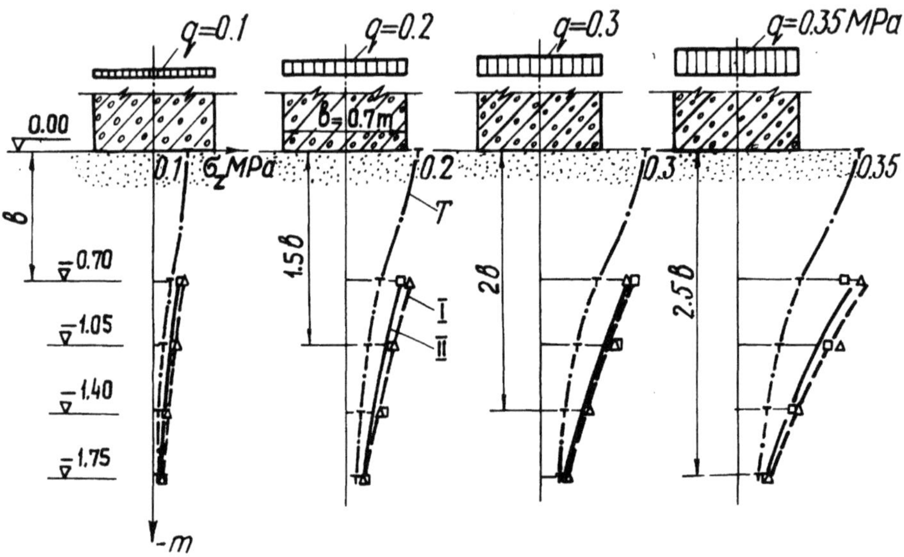

Figure 12-2. Soil pressure distribution in soil mass under stiff rectangular block 2.85 × 0.7 × 0.7 m; q = load increments; T = stress distribution calculated by classical theory of elasticity. I. Stresses obtained in first loading cycle; II. stresses obtained in second loading cycle.

12.2 Stiff Rectangular Concrete Block 2.85 × 0.7 × 0.7 m

This block is shown in Figures 2-1 and 2-2. In this investigation the block was tested placed on a foundation soil composed of sand with prevailing grain sizes as follows: 0.5 to 0.25 mm (52.9%) and 0.25 to 0.1 mm (39.9%). The average initial void ratio was $e = 0.64$ at a water content $w = 2.6$ to 2.8%. In this series of tests the soil pressure was measured by stiff flat pressure cells with hydraulic multiplicators (Figures 1-4 and 1-5). This time the cells were placed in the soil mass at elevations –0.7, −1.05, −1.4, and −1.75 m. The load was applied in two cycles in increments of 0.1, 0.2, 0.3, and 0.35 MPa.

The obtained results are shown in Figure 12-2. Again, as is seen from this figure, the actual measured stressed appear to be substantially larger than those obtained by the classical theory of elasticity (Timoshenko 1934; Hetinyi 1946; Gorbunov-Possadov and Malikova 1973). This is particularly true for larger applied loads.

12.3 Stiff Round Concrete Block 1.1 m in Diameter

This block is shown in Figure 2-8. In this investigation it was tested on a sandy foundation with prevailing grain sizes as follows: 0.5 to 0.25 min (65.8%) and

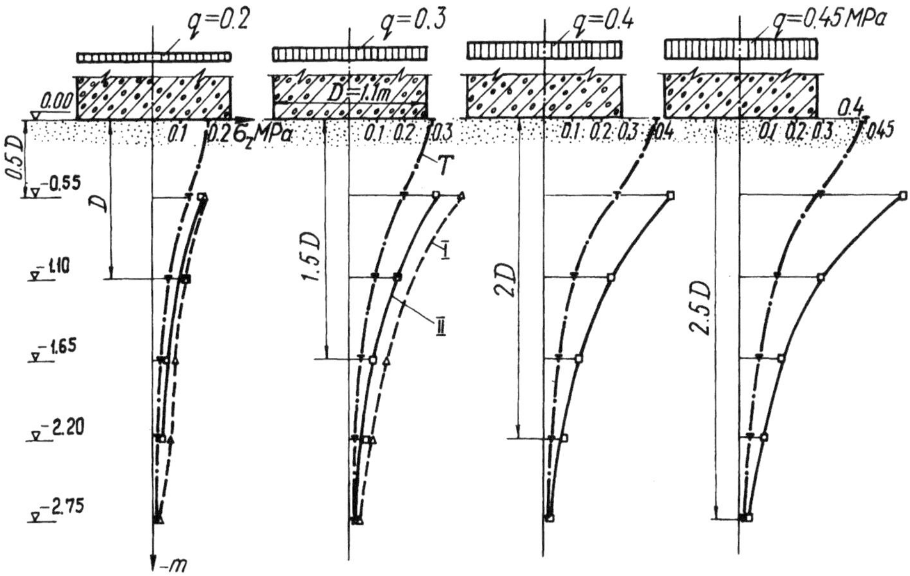

Figure 12-3. Soil pressure distribution in soil mass under stiff round block 1.1 m in diameter. T, I, and II. stresses obtained by classical theory of elasticity and during first and second loading cycles, respectively.

0.25 to 0.1 mm (23.9%). The initial density of the sand was 16.4 to 16.7 kN/m^3 at a water content w = 1 to 3%. The soil pressure in the soil mass was measured by thin, stiff, flat, large-diameter (d = 140 and 180 mm) pressure cells with hydraulic multiplicators and double measuring systems, one was based on the use of bonded toil strain gages and the other on the vibrating wire principle. These instruments are described in Chapter 7. This time the load was applied in increments of 0.2, 0.3, 0.4, and 0.45 MPa. The results of this investigation are illustrated in Figure 12-3.

Not surprisingly, in this case also the actual stresses in the mass of soil are substantially larger than those obtained from the classical theory of elasticity. Stresses in the mass of soil in excess of those obtained by the theory of elasticity were also measured by Turnbull et al. (1961); this was consistent in both sandy and loamy foundations.

12.4 Conclusions

The following conclusion can be drawn from the experiments described in this chapter.

1. The stresses in the mass of soil at the central vertical axis of a stiff foundation block resulting from a comparatively fast load application

exceed the theoretical values obtained by the classical theory of elasticity.

2. Unloading and repeated loadings of foundation soil via stiff blocks of different configurations lead to some decrease in the stress values along the central axis of these blocks.

References

Abramov, L. T., Kryzhanovsky, I. M. and Petrova, A. G. 1968. Investigation of Stress Distribution in Statistically Loaded Soils. SM and FE, N°6.

Abramson, L. W., and Green, G. E. 1985. *Reliability of Strain Gages and Load Cells to Geotechnical Engineering Application.* Reliability of Geotechnical Instrumentation. Trans. Res. Record, No. 1004.

Alberro, J., and Borbon, J. Z. 1985. *Testing of Earth Pressure Cells.* Behavior of Dams Built in Mexico (1974–1984) Vol. II, Commission Federal de Electricidad, Contribution to 15th ICOLD, Lausanne, Switzerland.

Alpan, I. 1967. *The Empirical Evaluation of the Coefficients K_0 and K_{or}.* Soil and Foundations (Japanese Society of Soil Mechanics and Foundation Engineering), Vol VII, No. 1, January.

Arthur, K. 1970. *Transducer Measurements.* Measurement Concept Series, Tektotronix, Inc., Beaverton, Oregon.

Baranov, D. S. 1959. *Measuring Instruments; Methods and Some Results of Investigation of Pressure Distribution in Sandy Soils.* Scientific Information, V. 7. TSNIISK, Moscow (in Russian).

Baranov, D. S. 1962. *Selection of the Basic Parameters of Soil Pressure Cells Based on Minimum Distortion of Measured Pressures.* Development of a method of wire tensometry for investigations of building structures. Transactions of TSNIISK, V. 14, pp. 40–84. Gosstroyizdat, Moscow (in Russian).

Baranov, D. S. 1982. *Guidelines on Application of Direct Methods of Pressure Measurement in Loose Materials and Soils.* Stroyizdat, Moscow (in Russian).

Baranov, D. S. 1987. To the Problem of Metrological Certification of Transducers Designed to Measure the Pressure in Granular Media. Transactions of TSNIISK, Moscow.

Baranov, D. S.; Sidorchuk, V. F.; and Karamzin, V. E. 1971a. *Meteorological Test of TSNIISK Sensors and Their Results.* Wire tensometry instruments for investigations of building structures, pp. 52–85. Stroyizdat, Moscow (in Russian).

Baranov, D. S.; Sidorchuk, V. F.; Karamzin, V. E.; and Zinovyev, R. K. 1971b. *Results of Methodical Tests of Sensors in Various Media and Materials.* Wire tensometry instruments for investigations of building structures, pp. 85–113. Stroyizdat, Moscow (in Russian).

Barker, W. R., and Reese, L. C. 1969. *Instrumentation for Measurement of Axial Load in Drilled Shafts.* Res. Rep. 89-6. Center for Highway Research, University of Texas, Austin, Texas.

Bhatia, S. K. and Blaney, G. W. (editors). 1991. *Recent Advances in Instrumentation, Data Acquisition and Testing in Soil Dynamics.* ASCE Proceedings of Sessions Sponsored by the Geotechnical Engineering Division, Orlando, Florida.

Binnie, Genesis. M. et al. 1967. *Mangla, Part 1, Engineering of Mangla.* Proceedings of the Institution of Civil Engineers, London, Vol. 38, November.

Bishop, A. W. 1958. *Test Requirements for Measuring the Coefficient of Earth Pressure at Rest.* Proceedings of Conference on Earth Pressure Problems, V. I, Brussels, Belgium pp. 2–14.

Bordes, J-L., and Debreuille, P-J. 1985. *Some Facts About Long-Term Reliability of Vibrating Wire Instruments.* Reliability of Geotechnical Instrumentation, Trans. Res. Record, No. 1004.

Borodachev, N. M. 1972. *On Control of Reactive Pressure Distribution Under the Footing.* Strength of Materials and Theory of Constructions, Issue No. 18, pp. 8–11. Budivelnyk, Kiev (in Russian).

Boussinesq, J. 1885. *Applications des potentials a l'etude de l'equilibre et du monvement des solides elastiques.* Gauthier-Villars, Paris, France.

Bozozuk, M. 1970. *Field Instrumentation of Soil.* Proceedings of the Conference on Design and Installation of Pile Foundations and Cellular Structures, Lehigh University, Bethlehem, Pennsylvania.

Bozozuk, M. 1972. *The Gloucester Test Fill.* Ph.D. Thesis, Purdue University, Lafayette, Indiana.

Brooker, E. W., and Ireland, H. O. 1965. *Earth Pressure at Rest Related to Stress History.* Canadian Geotechnical Journal, Vol. II, No. 1, February.

Brown, S. F. 1977. *State-of-the-Art Report on Field Instrumentation for Pavement Experiments.* Multiple Aspects of Soil Mechanics. Trans. Res. Record, No. 640.

Budin, A. Ya., and Nerpin, S. V. 1962. *Electrocontact Gage with Balanced Membrane.* Proceedings LIVT, Vol. XXXVIII, Leningrad, USSR (in Russian).

Burmistrov, M. A. 1961. *Soil Pressure on Foundation Soil and Walls of the Navigation Lock.* Laboratory and full-scale observations in Volga Lenin hydroproject. Transactions of GIDROPROJECT, V. 5, pp. 159–165. GIDPROPROJECT Publ. House, Moscow (in Russian).

Carlson, R. W. 1975. *Manual for the Use of Strain Meters and Other Instruments for Embedment in Concrete Structures, 4th ed.* Carlson Instruments, Campbell, California.

Carlson, R. W. 1978. *Performance Tests on Stress Meters.* International Water Power and Dam Construction, V. 30, No. 4, April, London.

Carlson, R. W. 1984. Personal communication with J. Dunnicliff (Dunnicliff 1988).

Casagrande, A., and Wilson, S. D. 1953. *Stability of Bulkhead at Boston Army Base.* Report to the U.S. Army Corps of Engineers, August.

Casagrande, L. 1973. *Comments on Conventional Design of Retaining Structures.* ASCE Journal of the Soil Mechanics and Foundation Division, V. 99, No. SM2, February.

Cerni, R. H., and Foster, L. E. 1965. *Instrumentation for Engineering Measurements.* John Wiley & Sons, New York.

Chernyavski, V. L.; Kozierev, V. S.; and Mendelev, Ya. I. 1987. *Investigation of Soils Interaction of Nuclear Powerplant with Power Unit VVER-1000.* Constructions of nuclear powerplants. Express-information of CNTI, Minenergo, USSR, Issue 7, Moscow (in Russian).

Chernyshova, K. I. 1963. *Equipment for Measurements of Lateral Earth Pressures on Retaining Wall.* Reports of the Ukrainian Academy of Sciences, No. 6, pp. 744–748. Kiev (in Ukrainian).

Clayton, C. R. I.; Milititsky, J.; and Woods, R. I. 1993. *Earth Pressure and Earth Retaining Structures.* Blackie Academic and Professional, C&H, London.

Considine, D. M. 1971. *Encyclopedia of Instrumentation and Control.* McGraw-Hill, New York.

Corbett, D. A.; Coyle, H. M.; Bartoskewitz, R. E.; and Milberger, L. J. 1971. *Evaluation of Pressure Cells Used for Field Measurements of Lateral Earth Pressures on Retaining Walls.* Research Report 169-1, Texas Transportation Institute, Texas A&M University, College Station, Texas.

Davidenkov, N. N., 1933. Vibrating-wire method used for measurements of strain. GTTI, Moscow-Leningram (In Russian).

Darwin, G. H. 1883. *On the Horizontal Thrust of a Mass of Sand.* Minutes of the Proceedings of the Institution of Civil Engineers, V. LXXI, Paper No. 1904, London.

Dergachev, P. V. 1957. *On Soil Pressure Measurement.* Information materials, No. 6, pp. 3–22. VODGEO, Moscow (in Russian).

DiBiagio, E. 1977. *Field Instrumentation—A Geotechnical Tool.* Norwegian Geotechnical Institute, Publication No. 115.

Dreyer, H. 1977. *Long-Term Measurements in Rock Mechanics by Means of Maihak Vibrating Wire Instrumentation.* Proceedings of the International Symposium on Field Measurements in Rock Mechanics, V. 1, Zürich.

Dunnicliff, J. 1988. *Geotechnical Instrumentation for Monitoring Field Performance.* John Wiley & Sons, New York.

Eidelman, S. Ya. 1960a. *Full-Scale Investigations of Concrete Hydraulic Structures.* Gosenergoizdat, Moscow-Leningrad (in Russian).

Eidelman, S. Ya. 1960b. *Contact Stresses in Founding Soils of Concrete Structures at Kakhovka Hydroproject.* Izv, VNIIG, V. 63. Gosenergoizdat, Leningrad (in Russian).

Eidelman, S. Ya. 1964. *Preliminary Guidelines on Arrangement of Controlling and Measuring Devices Installed into Concrete Hydraulic Structures Carrying Out of Measurements and Data Processing.* Energia, Leningrad (in Russian).

Evdokimov, P. D. 1962. *Methods of Controlling Contact Stresses in Soils.* Proceedings of the Coordination Conference on Hydraulic Engineering, Issue No. 3, pp. 110–116. Gosenergoizdat, Moscow-Leningrad (in Russian).

Fedorov, I. V., and Malyshev, M. V. 1954. *On Lateral Earth Pressures in Sandy Soils.* Hydr. Constr., No. 6, pp. 18–22, Moscow.

Franklin, J. A. 1977. *Some Practical Considerations in the Planning of Field Instrumentation.* Proceedings of the International Symposium on Field Measurements in Rock Mechanics, K. Kovari (ed.) V. 1. Balkema, Rotterdam, The Netherlands.

Franz, G. 1958. Unmittelbare Spannungsmessung in Befon und Baugrund. Der Baningenieur, Jg. 33, No. 5.

Fröhlich, O. K. 1934. *Druckverteilung im Baugrunde mit besonderer Berücksichtigung der plastischen Erscheinungen.* Springer, Berlin.

Fumagalli, E. 1969. *Tests of Cohesionless Materials for Rockfill Dams.* Proceedings ASCE, V. 95, pp. 313–330, No. SMI.

Goldstein, M. N. 1982. *Mechanical Properties of Soils and Perfection of Methods of Soil Investigations.* Soil Mechanics and Foundation Engineering, No. 3, pp. 21–23. Moscow.

Gorbunov-Possadov, M. I., and Malikova, T. A. 1973. *Analysis of Structures on Elastic Foundation.* Stroyizdat, Moscow (in Russian).

Green, G. E. 1986. Personal communication with J. Dunnicliff (Dunnicliff, 1988).

Green, G. E.; Feldman, A. I.; and Thordareson, P. 1983. *A Pile Tip Load Cell for a Driven Concrete Pile.* Proceedings of International Symposium on Field Measurements in Geomechanics, V. 1, Zurich.

Guidelines for Determination of Contact Soil Pressure Using Pressure Cells SDKS. (VTU 163–66) 1967. Reklama, Kiev (in Russian).

Hadala, P. F. 1967. *The Effect of Placement Method on the Response of Soil Stress Gages.* Proceedings of International Symposium on Wave Propagation and Dynamic Properties of Earth Materials. The University of New Mexico Press, Albuquerque, New Mexico.

Hakman, P. A., and Buser, W. M. 1961. *Deflection and Stress Measurements on New Sheet Pile Bulkhead at Port Toledo, Ohio.* Presented at the October 19 ASCE Convention, held in New York, NY.

Hast, N. 1945. Measuring Stresses and Deformations in Soil Materials. Ingenjörsvetenskapsakad. Handl. No. 178, Stockholm.

Hawkes, I., and Bailey, W. V. 1973. *Design, Develop, Fabricate, Test and Demonstrate Permissible Low Cost Cylindrical Stress Gages and Associated Components Capable of Measuring Change in Stress as a Function of Time in Underground Coal Mines.* U.S. Department of the Interior, Bureau of Mines, Rep., Cont. No. HO220050.

Herceg, E. G. 1972. *Handbook of Measurement and Control.* Schaevitz Engineering, Inc. Pennsanken, New Jersey.

Hetinyi, M. 1946. *Beams on Elastic Foundations.* University of Michigan Press, Ann Arbor, Michigan.

Horwath, R. G. 1985. *Instrumentation for Load-Transfer in Socketed Pier Foundations.* Reliability of Geotechnical Instrumentation, Trans. Res. Record, No. 1004.

Hvorslev, M. J. 1976. *The Changeable Interaction Between Soils and Pressure Cells: Tests and Reviews at the Waterway Experiment Station, Technical Report No. 5-76-7.* U.S. Army Corps of Engineers, Waterway Experiment Station, Vicksburg, Mississippi.

ISRM. 1981. *Suggested Methods for Pressure Monitoring Using Hydraulic Cells.* Rock Characterization Testing and Monitoring, ISRM Suggested Methods. Pergamon Press, Oxford.

Jaky, J. 1944. *The Coefficient of Earth Pressure at Rest.* Journal of the Society of Hungarian Architects and Engineers.

Kallstenius, T., and Bergau, W. 1956. *Investigations of Soil Pressure Measuring by Means of Pressure Cells.* Royal Swedish Geotechnical Institute, Proc. No. 12. Stockholm.

Kennard, M. F.; Penman, D. M.; and Vaughan, P. R. 1967. *Stress and Strain Measurements in the Clay Core at Balderhead Dam.* Transaction of the 9th ICOLD, V. III, Istanbul, International Commission on Large Dams, Paris, France.

Kheifitz, V. Z. 1973. *Measurement of Soil Stresses.* Informenergo, Moscow (in Russian).

Kheifitz, V. Z., and Petrashev, I. R. 1973. *Sensor for Measurements of Soil Stresses.* Additional papers to the 8th ICSMFE, pp. 200–202. Stroyizdat, Moscow (in Russian).

Kheifitz, V. Z., and Radkevich, D. B. 1969. *Device for Calibration of Soil Pressure Sensors.* USSR Patent Certificate No. 247577. Bulletin No. 22. Moscow (in Russian).

Kolbuszewski, J. 1958. *Fundamental Factors Affecting Experimental Procedures Dealing with Pressure Distribution in Sands.* Proceedings of Conference on Earth Pressure Problems, V. I, pp. 71–81. Brussels.

Kravtzov, V. I., and Evdokimov, P. D. 1957. *To the Problem of Soil Contact Pressure Distribution at Interface with Stiff Footings.* Izv. VNIIG, V. 57, pp. 71–76. Gosenergoizdat, Moscow-Leningrad (in Russian).

Krivorotov, A. P. 1971. *On the Technique of Pressure Measurement in Soils.* Soil Mechanics and Foundation Engineering, No. 1, Moscow.

Kuznetsov, G. N., and Budko, M. N. 1961. Guidelines on measuring of deformations in models made from compatible materials by means of deformometers D-2 type and modular strain gages TB-1 type. VNIMI, Leningrad (in Russian).

Kyle, J. M., and Kapp, M. S. 1957. *A Flexible Bulkhead for New York Harbor.* Proceedings of the 4th International Conference on Soil Mechanics and Foundation Engineering. London, V. 2.

Kyowa's Civil Engineering Instruments. 1973. Soil Pressure Transducer Series BE. Kyowa Electronic Instruments Co. Ltd., Cat. No. BT/B., Japan.

Lambe, T. W. 1951. *Soil Testing for Engineers.* John Wiley & Sons, London.

Lambe, T. W., and Whitman, R. V. 1969. *Soil Mechanics.* John Wiley & Sons, New York.

Lazebnik, G. E., 1961a. *Thin Anchored Retaining Walls.* Proceedings of Conference of Young Scientists. Academy of Civil Engineering and Architecture pp. 142–147. Kiev (in Russian).

Lazebnik, G. E. 1961b. *Investigations of Thin Single-Anchored Retaining Walls and Recommendations for their Analysis.* In Proceedings of the Investigations of Stressed State of Hydrotechnical Constructions, pp. 115–129. Ukrainian Academy of Sciences, Kiev (in Ukrainian).

Lazebnik, G. E. 1961c. *Analysis of Sheet-Pile Retaining Walls.* River transport, No. 2, pp. 39–43, Moscow (in Russian).

Lazebnik, G. E. 1966. *Investigation of Contact Stresses at Interface of Concrete Block and Stress Distribution within Mass of Soil.* NIISP, unpublished report (in Russian).

Lazebnik, G. E. 1967. *Determining of Compression Characteristics of Cohesionless Soils.* Reklama, Kiev (in Russian).

Lazebnik, G. E. 1969. *Development, Testing and Improvement of Soil Pressure Cells for Full-Scale Investigations of Soil/Structure Interaction.* Investigations on founding soils, footings and soil mechanics, pp. 116–123. Budivelnyk, Kiev (in Russian).

Lazebnik, G. E. 1970a. *Experimental Investigations of "Orientation Effect" and Development of New Types of Soil Pressure Cells for Measurements of Soil Pressure in Mass of Soil.* Izv. VNIIG, V. 93, pp. 200–208. Energia, Leningrad (in Russian).

Lazebnik, G. E. 1970b. *Soil Pressure Cells for Measurement of Stresses in Mass of Soil in Full-Scale and Large-Scale Experiments.* Reklama, Kiev (in Russian).

Lazebnik, G. E. 1970c. *Full-Scale Investigation of Contact Stress Distribution at Underside of Foundation Slabs of Multistorey Buildings.* Soil Mechanics and Foundation Engineering, No. 6. Moscow.

Lazebnik, G. E. 1973. *Founding Soil/Structure Interaction.* Proceedings 8th ICSMFE, V. 4.2, pp. 53–55. Moscow.

Lazebnik, G. E. 1974a. *Pressure Cell for Pressure Measurements in Mass of Loose Media.* USSR Patent Certificate No. 414503. Bulletin No. 5. Moscow (in Russian).

Lazebnik, G. E. 1974b. *Measurements of Contact Soil Pressure at Prefabricated Reinforced Concrete Reservoir and Displacement of Reservoir's Cover.* Experimental Investigations of Structures (Methods, Equipment and Devices), Issue No. 89, pp. 15–20. CNIIS publisher, Moscow (in Russian).

Lazebnik, G. E. 1975. *Device for Calibration of Soil Pressure Cells.* USSR Patent Certificate No. 473921. Bulletin No. 22, Moscow (in Russian).

Lazebnik, G. E. 1980. *Development and Testing of Soil Pressure Cells for Measurement of Static and Dynamic Stresses in a Single Experimental Point.* Special construction work. Proc. VNIIGS, pp. 112–118. Leningrad (in Russian).

Lazebnik, G. E. 1982. *Contact Soil Pressure Cells with Extended Range of Measurements.* Express-information of CBNTI, USSR Minvodkhoz, Series No. 10, Issue No. 3, Moscow (in Russian).

Lazebnik, G. E. and Smirnov, A. A. 1968. Device for Simultaneous Calibration of Several Contact Pressure Cells. USSR Patent No. 214192. Bulletin No. 11, Moscow (In Russian).

Lazebnik, G. E. and Tivolap, P. G. 1974. Some Factors that Affect the Magnitude of the Lateral Pressure "At Rest" of the Granular. Izv.VNIIG v. 105, pp. 64–72, Energia, Leningrad (In Russian).

Lazebnik, G. E. and Sivolap, P. G. 1977. Devices for Determining the Modulus of Deformation and Coefficient of Lateral Pressure of Granular Soils "At Rest". Proceedings VNIIGS, Issue No. 45, Leningrad (In Russian).

Lazebnik, G. E., and Sivolap, P. G. 1981. *Dependence of the Coefficient of Lateral Pressure of Sand "At Rest" on the Configuration of Calibration Chamber.* Founding Soils and Footings. Republican interdisciplinary collection of scientific and engineering papers, Issue No. 14, pp. 48–51. Budivelnyk, Kiev (in Russian).

Lazebnik, G. E., and Smirnov, A. A. 1964. *Measurement of Stresses Under a Rigid Foundation Block.* Soil Mechanics and Foundation Engineering, No. 2, Moscow.

Lazebnik, G. E., and Smirnov, A. A. 1972. *Experimental Investigation of Errors in Contact Stress Measurement by Soil Pressure Cells.* Izv. VNIIG, V. 98, pp. 93–102. Energia, Leningrad (in Russian).

Lazebnik, G. E.; Smirnov, A. A.; and Ivanov, D. G. 1973. *Set of Instruments and Equipment for Measuring of Soil Pressures.* Soil Mechanics and Foundation Engineering, No. 2.

Lazebnik, G. E.; Smirnov, A. A.; Simakov, V. I.; and Levadny, Ya. M. 1966. *On Method of Measuring Stresses in Soil under Footings.* Soil Mechanics and Foundation Engineering, No. 5, Moscow.

Lion, K. S. 1959. *Instrumentation in Scientific Research, Electrical Input Transducers.* McGraw-Hill, New York.

Lipovetskaya, T. F. 1953. *Experimental Investigations of Stress Distribution Under the Rigid Slabs Founded on Sandy Foundations.* Izv. VNIIG, V. 49, pp. 54–64. Gosenergoizdat, Leningrad (in Russian).

Loh, Y. C. 1952. *Internal Stress Gages for Cementitious Materials.* Proceedings of the Society for Experimental Stress Analysis, V. 11, No. 2.

Lomakin, V. A., and Sheinin, V. I. 1970. *Statistical Characteristics of Stress Fields in Random Heterogeneous Elastic Plane.* Mechanics of Rigid Body, No. 4, pp. 124–130. Moscow (in Russian).

Muhs, H. 1947. *Erddruckmessungen an einer 24 m hohen starren Wand.* Bauplanung und Bautechnik, V. 1, No. 1, Berlin.

Murzenko, Yu. N. 1965. *Distribution of Contact Stresses in Sand on Contact with Stiff Footing.* Soil Mechanics and Foundation Engineering, No. 2, Moscow.

Norton, H. N. 1969. *Handbook of Transducers for Electronic Measuring Systems.* Prentice-Hall, Englewood Cliffs, New Jersey.

Nowack, F., and Gartung, E. 1983. *Instrumentation of Cast-in-Place Piles for Vertical and Horizontal Load Testing.* Proceedings of International Symposium on Field Measurements in Geomechanics, V. 1, Zurich.

Ohde, I. 1938. *Zur Theorie des Erddruckes unter besonderer Berücksichtigung der Erddruckverteilung.* Die Bautechnik, No. 10/11, 13, 19, 25, 37, 42, 53–54. Berlin.

O'Rourke, J. E. 1978. *Soil Stress Measurement Experiences.* ASCE Journal Geotechnical Engineering Division, V. 104, No. GT12, December.

Peattie, K. R., and Sparrow, R. W. 1954. *The Fundamental Action of Earth Pressure Cells.* Journal of Mechanics and Physics of Solids, No. 2, pp. 141–155.

Peck, R. B. 1969. *Advantages and Limitations of the Observational Method in Applied Soil Mechanics.* Geotechnique, V. 19, No. 2.

Peck, R. B., 1988. Foreword to *Geotechnical Instrumentation for Monitoring Field Performance,* by Dunnicliff, J., 1988. John Wiley & Sons, New York.

Perry, C. C., and Lissner, H. R. 1962. *The Strain Gage Primer.* McGraw-Hill, New York.

Plantema, I. G. 1953. *A Soil Pressure Cell and Calibration Equipment.* Proceedings 3rd Congress ICSMFE, V. 1, pp. 283–288, Switzerland.

Pokrovsky, G. I. 1937. *Theory of Impact and Explosion, Part I.* VIA RKKA, Moscow (in Russian).

Prensky, S. D. 1963. *Electronic Instrumentation.* Prentice-Hall, Englewood Cliffs, New Jersey.

Price, G., and Wardle, I. F. 1983. *Recent Developments in Pile/Soil Instrumentation Systems.* Proceedings of International Symposium on Field Measurements in Geomechanics, V. 1, Zurich.

Reese, L. D.; Brown, J. C.; and Dalrymple. 1968. *Instrumentation for Measurements of Lateral Earth Pressure in Drilled Shafts.* Research Report 89-2, Center for Highway Research University of Texas, Austin, Texas.

Renski, A. B.; Baranov, D. S.; and Makarov, R. A. 1977. *Measurements of Strain in Structures and Materials.* Stroyizdat, Moscow (in Russian).

Rowe, P. W. 1952. *Anchored Sheet-Pile Walls.* Proceedings Institute of Civil Engineers, V. 1, pp. 27–70, No. 1.

Rowe, P. W. 1954. *A Stress-Strain Theory for Cohesionless Soil with Applications to Earth Pressure "At Rest" and Moving Walls.* Geotechnique, N.Y., No. 2.

Rowe, P. W. 1958. Measurement of Sheet Pile Walls Driven into Clay. Proc. of the Conference '58 on Earth Pressure Problems, V. 2, Brussels.*

Sammal, O. Yu. 1980. *Stress in Concrete and Prediction of Technical Resources in Concrete and Steel Concrete Structures and Constructions.* Tallinn: Valgus (in Russian).

Selig, E. T. 1964. *A Review of Stress and Strain Measurement in Soil.* In Proceedings of the Symposium on Soil-Structure Interaction. University of Arizona, Tucson, Arizona.

Selig, E. T. 1980. *Soil Stress Gage Calibration.* Geotechnical Testing Journal, ASTM, V. 3, No. 4, December.

Sivolap, P. G., and Ivanov, D. G. 1977. *Investigation of Stressed State Condition of Cohesionless Soil in Calibration Device and Soil Interaction with Soil Pressure Cells.* Special construction work, Proceedings VNIIGS, Issue No. 45, pp. 131–136, Leningrad (in Russian).

Smirnov, A. A. 1972. *Development and Testing the Pressure Cells Designed to Measure Soil Pressures on Contact with Structures.* Proceedings of the 2nd International Mine Surveying Conference, V. 3, Budapest.

Sorotchan, E. A.; Troitzky, G. M.; Klepikov, S. N.; Lazebnik, G. E.,; Vinokourov, E. F.; and Bykhovtsev, V. E. 1981. *Soil/Foundation Interaction.* Proceedings 10th ICSMFE, V. 2, pp. 259–262. Stockholm.

Spitzer, P., and Howarth, B. 1972. *Principles of Modern Instrumentation.* Holt, Rinehart and Winston, New York.

State of California, 1968. *Evaluation of Commercial Soil Pressure Cells.* Research Report No. M&R 636342. Materials and Research Dept., State of California, Sacramento.

State of California, 1971. *Field Performance Evaluation of Soil Pressure Cells.* Research Report No. M&R 632954-1. Material and Research Dept., State of California, Sacramento.

Stein, P. K. 1964. *Measurement Engineering,* V. 1, Basic Principles. Stein Engineering Services, Phoenix, Arizona.

Tarikuliev, Z. Ya., and Murzenko, Yu. N. 1969. *Investigation of Contact Stresses Distribution at Square Slab/Sand Interface.* Experimental investigations of structures. Proceedings of the 2nd Symposium, RIO NPI, pp. 94–99. Novocherkassk (in Russian).

Terzaghi, K. 1920. *Old Earth Pressure Theories and New Test Results.* ENR, V. 85, pp. 632–637.

Terzaghi, K. 1925. *Erdbaumechanik auf Bodenphysikalischer Grundlage.* Leipzig-Wien.

Terzaghi, K. 1934. *Large Retaining Wall Tests.* Engineering News Record, Feb 1, V. 112.

Terzaghi, K. 1953. Anchord Bulkheads. Proceedings ASCE v. 79. Sept. No. 262.

Tharp, D. 1986. Personal communication with J. Dunnicliff (in Dunnicliff 1988).

Thomas, H. S. H. 1966. *The Measurement of Strain in Tunnel Lingins Using the Vibrating-Wire Technique.* Strain, V. 2, July.

Thomas, H. S. H., and Ward, W. H. 1969. *The Design, Construction and Performance of a Vibrating Wire Earth Pressure Cell.* Geotechnique, V. 19, No. 1.

Timoshenko, S. 1934. *Theory of Elasticity.* McGraw-Hill, New York.

Trollope, D. H., and Lee, I. K. 1961. *The Measurement of Soil Pressures.* Proceedings 5th ICSMFE, V. 2, pp. 493–499. Dunod, Paris.

Tschebotarioff, G. P. 1948. *Large-Scale Model Earth Pressure Tests of Flexible Bulkheads.* Proceedings ASCE, V. 74, No. I, pp. 9–48.

Tschebotarioff, G. P. 1973. *Foundations, Retaining and Earth Structures.* McGraw-Hill, New York.

Tschebotarioff, G. P. 1973. *Foundations, Retaining and Earth Structures* (2nd ed.). McGraw-Hill, New York.

Tsinker, G. P. 1997. *Handbook of Port and Harbor Engineering.* Geotechnical and Structural Aspects. Chapman & Hall, New York.

Turnbull, M. J.; Maxwell, A. A.; and Ahlvin, R. G. 1961. *Stresses and Deflections in Homogeneous Soil Masses.* Proceedings 5th ICSMFE, V. 2, Dunod, Paris.

Uff, J. F. 1970. *In Situ Measurements of Earth Pressure for a Quay Wall at Seaforth, Liverpool, UK.* Proceedings of the Conference on In Situ Investigations in Soil and Rock, British Geotechnical Society, London.

Uginchus, A. A., and Bombchinsky, V. P. 1954. *Instrumentation of Hydraulic Structures.* Gosstroyizdat, Moscow (in Russian).

Weiler, W. A., and Kulhawy, F. H. 1978. *Behavior of Stress Cells in Soils.* Report 78-2, School of Civil and Environmental Engineering, Cornell University, Ithaca, New York.

Weiler, W. A., and Kulhawy, F. H. 1982. *Factors Affecting Stress Cell Measurement in Soil.* ASCE Journal of Geotechnical Engineering Division, V. 108, No. GT12, December.

Wilson, S. D. 1982. *Instrumentation and Performance of Embankment Dams.* Lecture Notes, Nigerian Geotechnical Association, The Nigerian Society of Engineers, Lagos, Nigeria.

Wilson, S. D. 1984. Personal communication with J. Dunnicliff (in Dunnicliff, 1988).

Wnuk, S. P., Jr. 1981. *On the Use of Bonded Weldable Stain Gages for Field Measurements in Hostile Environments.* Proceedings of International Conference on Measurements in Hostile Environments, British Society for Strain Measurements, Edinburgh, UK.

Wolf, S. 1973. *Guide to Electronic Measurements and Laboratory Practice.* Prentice-Hall, Englewood Cliffs, New Jersey.

Index

Accuracy, of soil pressure cell, 6
Acrylic inserts, in nonstiff soil pressure cells, 88, 89
Active wedge, and backfill for sheetpile walls, 198
Air pressure, and calibration chambers, 38
Aluminum alloy, and single-anchored sheetpile wall, 186–87
Anchor points, of sheetpile walls, 193, 195, 196
Angle of internal friction, and vertical stresses in sand, 84
Apartment building (Kiev, Ukraine), 154–58
Argillite, 141
Aspect ratio, of soil pressure cell, 5

Backfill
  sheetpile walls, 195, 197–98, 199–203
  variability in composition or density, 73–74
Baranov formula, 3, 93–94
Baranov's pressure cell, 6
Bending moment diagram, for sheetpile walls, 192–93, 196
Benkelman-Lancaster device, 39
Bonded electrical resistance strain gages, 62–63
Bonded foil resistance strain gage transducer, 20, 21
Bonded foil strain gages, 58, 207
Bonded wire strain gage, 57–58
Boussinesq solution, 204
Bridge abutment, and lateral soil pressure, 74, 76
Bus parking station, cable roof system for, 135–38

Cable roof system, round concrete foundation for cylindrical concrete shaft to support, 135–38
Calibration chambers
  large soil pressure measurement devices, 13
  performance evaluation of improved devices, 45–56
  review of available instruments, 38–45
Calibration curves
  flat soil pressure cell, 31
  membrane-type soil pressure cells, 4
  nonrigid soil pressure cell, 15, 16
  orientation of soil pressure cell, 11
  PDGS–1 soil pressure cell, 9, 10
  rigid symmetrical soil pressure cells, 18
  soil pressure cell SDKS–7.1, 73
  vibrating wire transducer, 20, 21
Calibration device SCC–475–0.5, 47, 48, 49, 50, 51, 52–54
Carlson Instruments, 63
Carlson's pressure cell, 6, 7, 8
Case histories, of field monitoring investigations
  cellular square footing on granular foundation soil, 138–40
  cellular square footing on hard foundation soil, 141–44
  foundation slab to support loads from turbine and generator, 148–52
  foundation of tall building, 144–48
  round concrete foundation for cylindrical concrete shaft for cable roof system, 135–38
Cell placement, effects of on soil pressure cells, 15–16

Cell size, and scatter in soil pressure measurement, 73–78
Cellular square footing
granular foundation soils, 138–40
hard foundation soil, 141–44
Clay, lateral earth pressure and plasticity index of, 166
Cohesionless soils, nonstiff pressure cells and compression of, 83–88
Cohesive soils, perforated stiff foundation system on, 153–54, 155
Compressibility and compression
nonstiff pressure cells and properties of cohesionless soils, 83–88
variability of in foundation soil or backfill, 73–74
Concrete water storage reservoir, 158–60
Construction
contractors and field monitoring of soil-structure interaction, 114–15
sheetpile walls, 195, 197–98
soil pressure readings at end of pre-subscribed stages, 136
Contact earth pressure cells, 2
Contact soil pressure cell SDKS–1–0.5, 65–67
Contact soil pressure cell SDKS–6VMR, 67–70
Contact soil pressure cell SDKS–7, 70–72
Contact soil pressure measurement
calibration, 50
convention approach and experimental verification, 23–30
development of pressure cells, 59–70
Contact stresses, between stiff foundation block and foundation soil, 126–34
Contractors, and field monitoring of soil-structure interaction, 114–15
Cracked tuff, 141
Cylindrical concrete shaft, for cable roof system, 135–38

Dedicated contact pressure cells, 37
Deflections, at centers of pressure-bearing membrane of membrane-type contact pressure cells, 108
Deformability, of contact pressure cells, 60
Deformation, of soil. *See also* Lateral pressure; Vertical pressure
calibration devices, 53
contact stresses with foundation soil, 129, 131
methodology of soil pressure measurement, 2–4
nonstiff pressure cells, 87
uniaxial conditions and calibration of soil pressure cells, 39
Depth, in soil
performance of pressure-bearing membranes, 89–92
soil pressure measurement instruments and optimal, 29
Diaphragm-type pressure cells, 92
Direct method, of soil pressure measurement, 1–2
Distance, between soil pressure cells and instrument performance, 88–89
Distribution, of soil pressure under stiff foundations, 130, 134, 205 . *See also* Stress distribution
Double-cylinder contact soil pressure cell, 63
Dredging, and backfill for sheetpile walls, 195, 197–98
Dynamic soil pressure, cell for measurement of, 19–22

Elasticity, theory of
lateral earth pressure at rest, 165
vertical stress under stiff foundation, 205, 207
Electrical resistance strain gages, 57–59
Electric transducers, in soil pressure cells, 57–59
Electroacoustic transducers. *See* Vibrating wire transducers
Embedment earth pressure cells, 2
Energodar nuclear power station (Ukraine), 138–40, 142, 143
Error. *See also* Nonstiff pressure cells
definition of in context, 59–60
formula for relative in soil pressure measurement, 2

Field monitoring programs, of soil/structure interaction
cellular square footing on hard foundation soil, 141–44
cellular square footing on granular foundation soil, 138–40
foundation slab to support loads from turbine and generator, 148–52
foundation of tall building, 144–48
interaction between adjacent pressure cells in soil mass, 123–25

layout and installation of soil pressure cells, 112–18
misalignment of soil pressure cell contact surface, 118–21, 122
round concrete foundation for cylindrical concrete shaft for cable roof system, 135–38
soil grain size and cell performance, 121–23
Fine crushed stone, and grain size distribution, 177
Finite element analysis, 113
Flat membrane pressure cells, 153
Flexibility index, for foundation slab, 151
Flexural deformation, of nonstiff pressure cells, 80
Foundation geometry, and optimal depth for soil pressure measurement devices, 29
Foundation slab, to support loads from turbine and generator, 148–52
Foundation soils. *See also* Loam; Sand
anchor point for sheetpile walls, 193
cellular square footing on granular, 138–40
cellular square footing on hard, 141–44
contact stresses at interface with stiff foundation block, 126–34
nonuniform densification and shape of soil reaction diagram, 160–64
solid concrete slab on nonhomogeneous, 154–58
variability in composition or density, 73–74

Generator, foundation slab for, 148–52
Grain size, of soil
cell performance in field monitoring programs, 121–23
experiments with nonstiff pressure cells, 84
fine crushed stone, 177
Granular composition, of soil
foundation soil underlying structure of water reservoir, 158–59
lateral earth pressure, 166–70
variability of in foundation soil or backfill, 73–74
Gravity-type retaining walls, and lateral soil pressure, 74, 75
Ground water table, 141, 156
Guidelines for Determination of Contact Soil Pressure Using Pressure Cells SDKS (1967), 77

Hard foundation soil, cellular square footing on, 141–44
Hydraulic multiplicators
contact soil pressure measurement, 26, 37
lateral soil pressure, 32
measurement methodology, 6–9
nonstiff soil pressure cells, 79

Indirect method, of soil pressure measurement, 1–2
Installation. *See also* Layout; Placement
contact pressure cells on retaining wall, 202
large soil pressure cell, 14
soil pressure cells for field monitoring program, 112–18

Kheifitz formula, 3
Khmelnitsky nuclear power station (Ukraine), 141–44, 145
Kyowa Company (Japan), 74

Large calibration device LCC–1000–1, 48–49, 50
Large-size pistonless odometer, 44–45
Lateral soil pressure. *See also* Deformation
background information, 165–66
bridge abutment, 74, 76
formula, 183
granular soil mass, 166–70
gravity-type retaining walls, 74, 75
readings of soil pressure cells of different designs, 33
sheetpile walls, 192, 202, 203
stiff walls and coefficient of, 170–82
Layout. *See also* Installation; Placement
contact soil pressure measuring instruments in model tests, 130, 132
soil pressure cells in field monitoring program, 112–18
Loam
foundation soil for water reservoir, 158–59
perforated stiff foundation system, 153

Measurement, of soil pressure. *See also* Soil pressure cells
calibration chambers and calibration methods, 38–56
case histories of field investigations, 135–52
contact soil pressure measurement techniques, 23–37
contact stresses between stiff foundation block and foundation soil, 126–34
experiments on in situ monitoring, 112–25
lateral soil pressure at rest, 165–83

methodological issues, 1–22
nonstiff soil pressure cells and errors, 79–111
retaining walls, 184–203
stiff contact pressure cells, 57–78
stiff foundations on cohesive and nonhomogeneous soils, 153–64
vertical stress at center of stiff slab, 204–208
Membrane-type soil pressure cells
calibration, 4
nonstiff soil pressure cells, 79, 80, 91, 92, 94–96
Misalignment, of soil pressure cell surface with surface of structure, 118–21, 122
Model tests, of contact stresses between stiff foundation block and foundation soil, 126–34
Monitoring programs. *See* Field monitoring programs
Multicell large-diameter soil pressure measurement device, 82
Mushroom type, of nonstiff pressure cells, 80

Nonhomogeneous foundation soil, solid concrete slab constructed on, 154–58
Nonrigid soil pressure cell, 15, 16
Non-shift contact soil pressure cell, with vibrating wire transducer, 109–10
Nonstiff soil pressure cells, and error in measurement of soil contact pressure
background information, 79–80
compression and changes in properties of cohesionless soils, 83–88
depth of soil above pressure-bearing membrane and instrument performance, 89–92
distance between adjacent cells and instrument performance, 88–89
instruments and devices used in experiments, 80–82
interaction between adjacent cells, 124
results of experiments, 92–99
soil stress state investigation at contact surface, 99–104
systematic relative error in stress measurements, 106–11
Nonsymmetrical soil pressure cells, 16–17
Nonuniform densification, and shape of soil reaction diagram, 160–64
Nonuniformity, of soil and contact pressure cells, 73–78
Nonyielding retaining wall, soil pressure against, 198–203
Nuclear power stations, 138–40, 141–44

Odometer tests, 44–45, 165–66
Orientation effect, on performance and design of soil pressure cells, 9–18

Parallel displacement, of nonstiff pressure cells, 80, 109
PDGS-type soil pressure cell, 7–18
Perforated stiff foundation system, on cohesive soils, 153–54, 155
Pistonless odometer, 44–45
Placement, of soil pressure measurement instruments. *See also* Installation; Layout
calibration chambers, 40
optimal depth, 29
Plasticity index, clay and lateral earth pressure, 166
Plastic zones, and readings of soil pressure cells, 37
Precision, of soil pressure cells, 6
Pressure-bearing membranes, and depth of soil above nonstiff pressure cells, 89–92
Pumphouse, and soil pressure on retaining wall, 198–203

Reinforced concrete blocks, geometry of, 24
Retaining walls, soil pressure on
background information, 184–85
nonyielding walls, 198–203
single-anchored sheetpile wall, 185–98
Rigid symmetrical soil pressure cells
calibration curves, 17, 18
orientation effect, 17
Rigid test block, and contact soil pressure measurement, 36
Round concrete foundation, for cylindrical concrete shaft for cable roof system, 135–38
Round stiff concrete block, and contact stress with foundation soil, 127–28, 129, 130
Rowe formula, modified, 187
Rubber ring, effects of on soil pressure cells, 124–25

Sands. *See also* Foundation soils
lateral earth pressure, 166–70

nonstiff pressure cells and compression, 83–88
nonuniformity of and contact pressure cells, 73–78
plastic zones and readings of soil pressure cells, 37
stress distribution in mass of loose, 104
wall movement and earth pressure, 184, 185
Sandstone, 141
Scatter, and effects of cell size on soil pressure measurement, 73–78
Semiconductor resistance strain gage, 58
Sensitivity, of soil pressure cell, 6, 125
Settlement, of test blocks for contact soil pressure measurement, 32, 35
Shearing strength, and vertical stresses in sand, 84
Shear zones. *See* Plastic zones
Sheetpile walls, soil pressure on single-anchored, 185–98
Siltstone, 141
Single-anchored sheetpile wall, soil pressure on, 185–98
Single-cell installation, and errors in measuring soil contact pressures, 89
Soil. *See* Clay; Foundation soils; Grain size; Granular composition; Loam; Sand
Soil contact pressure, errors in measurement of by nonstiff pressure cells
background information, 79–80
compression and changes in properties of cohesionless soils, 83–88
depth of soil above pressure-bearing membrane and instrument performance, 89–92
distance between adjacent cells and instrument performance, 88–89
instruments and devices used in experiments, 80–82
results of experiments, 92–99
soil stress investigation at contact surface, 99–104
systematic relative error in stress measurements, 106–11
Soil friction, and calibration chambers, 40, 46, 49–50, 51, 52
Soil pressure, measurement of. *See also* Soil pressure cells
calibration chambers and calibration methods, 38–56
case histories of field investigations, 135–52
contact soil pressure measurement techniques, 23–37
contact stresses at interface between stiff foundation block and foundation soil, 126–34
experiments on in situ monitoring, 112–25
lateral soil pressure at rest, 165–83
methodological issues, 1–22
nonstiff pressure cells and errors, 79–111
retaining walls, 184–203
stiff contact pressure cells, 57–78
stiff foundations on cohesive and nonhomogeneous soils, 153–64
vertical stress at center of stiff slab, 204–208
Soil pressure cells. *See also* Soil pressure, measurement of; specific types
calibration chambers and calibration methods, 38–56
PDGS-type, 7–18
static and dynamic soil pressures, 19–22
stiff contact type, 57–78
Soil reaction diagram, influence of nonuniform densification on shape of, 160–64
Soil stress state investigation, and nonstiff soil pressure cells, 99–104
Soil/structure interaction, experiments on in situ monitoring of, 112–25. *See also* Soil pressure, measurement of; Soil pressure cells
Solid concrete slab, on nonhomogeneous foundation soil, 154–58
Sound wire transducer (WT), 20, 21
Spring clamp, soil pressure cells and stiff cohesive soil, 118
Static soil pressure, cell for measurement of, 19–22
Stiff circular concrete block, 161–62
Stiff contact pressure cells
cell size and scatter, 73–78
development of contact pressure cells, 59–70
stress-strain measuring systems, 57–59
vibrating wire type transducer, 70–72
Stiff flat soil pressure cells
interaction between adjacent cells, 124
rubber rings and sensitivity, 125
Stiff foundation block, model tests of contact stresses with foundation soil, 126–34
Stiff foundations, on cohesive and nonhomogeneous soils

nonuniform densification and shape of soil reaction diagram, 160–64
perforated stiff foundation system, 153–54, 155
solid concrete slab, 154–58
vertical stress at central part of stiff slab, 204–208
water reservoir, 158–60
Stiffness ratio, of soil pressure cell, 5–6
Stiff rectangular concrete block
contact stresses with foundation soil, 128, 132–34, 162–64
vertical stress at central part of, 204–206
Stiff reinforced concrete blocks, and contact soil pressure measurement, 23–30, 32
Stiff round concrete block, and vertical stress, 206–207
Stiff walls, and coefficient of lateral soil pressure, 170–82
Strain gage transducer (SGT), 19–22
Stress deviator, and variation in soil pressure cell readings, 30–37
Stress distribution. *See also* Distribution
contact stresses between stiff foundation blocks and foundation soil, 132–34
contact surfaces of nonstiff soil pressure cell, 99–104
Stress-strain measuring systems, 57–59
Survivability rate, for soil pressure cells in field monitoring programs, 113
Symmetrical soil pressure cells, 16, 17

Technical Information Institute (Ukraine), 144–48
Telescopic rings, and calibration apparatus for soil pressure cells, 44
Test blocks, for contact soil pressure measurement, 23–30, 32–37
Thermopower generating station (Kostroma, Russia), 148–52
Triaxial cells, 166
Turbine, foundation slab for, 148–52

Unbonded electrical resistance strain gages, 63–64
Unbonded wire strain gage, 58, 63
Uniaxial soil deformation, and calibration of soil pressure cells, 39
Uniform symmetrical stress state, and calibration devices, 53, 55, 56
U.S. Army Corps of Engineers, calibration chamber for soil pressure cells, 40–41
U.S. Army Engineer Waterways Experimental Station, soil pressure cell, 7

Vertical stress. *See also* Deformation
central part of stiff slab, 204–208
loose medium sand and lateral stress coefficients, 181
nonstiff pressure cells, 83, 84
stress deviator and variation in soil pressure cell readings, 30–37
Vibrating wire transducers
long-term monitoring of soil pressure changes, 18, 19–22
non-shift contact soil pressure cell, 109–10
stiff contact pressure cells, 65, 70–72
stress-strain measuring systems, 58–59
vertical stress, 207

Wall friction, and calibration of soil pressure cells, 40, 41
Water jacket, for calibration chambers, 41, 42–43
Water reservoir, soil pressure from, 158–60
Weldable resistance strain gages, 58
Wheatstone bridge circuit, 58, 62

Zero readings, and vibrating wire transducers, 59

MIX
Papier aus verantwortungsvollen Quellen
Paper from responsible sources
FSC® C105338

If you have any concerns about our products,
you can contact us on
ProductSafety@springernature.com

In case Publisher is established outside the EU,
the EU authorized representative is:
Springer Nature Customer Service Center GmbH
Europaplatz 3, 69115 Heidelberg, Germany

Printed by Libri Plureos GmbH
in Hamburg, Germany